The
SEXTANT
HANDBOOK

The
SEXTANT
HANDBOOK

Adjustment, Repair, Use and History

By

Cmdr. **Bruce A. Bauer**, USN (ret.)

Second Edition

International Marine
Camden, Maine 04843

Published by International Marine, an imprint of TAB Books.
TAB Books is a division of McGraw-Hill, Inc.

10 9 8 7 6 5 4 3 2 1

Library of Congress Cataloging-in-Publication Data
Bauer, Bruce A., 1931–
    The sextant handbook : adjustment, repair, use, and history / by
  Bruce A. Bauer.—2nd ed.
      p.   cm.
    Includes bibliographical references (p.   ) and index.
    ISBN 0-87742-344-X
    1. Sextant.   I. Title.
VK583.B38   1992
522′.4—dc20   92-2545
                                               CIP

Questions regarding the content of this book should be addressed to:

International Marine Publishing
P.O. Box 220
Camden, ME 04843

Printed by R.R. Donnelley
Line drawings by Nancy McCoy Bauer.
All photographs by author unless otherwise noted.

Dedication

To the wonderful girls in my life—mother Margaret, wife Nancy, sister Darlene, and daughters Lisa and Julie—who have waited and worried while I went to sea, welcomed me back lovingly, and been wise enough to let me go again and again without lament, appreciating, if not understanding, its necessity for me.

Acknowledgments

This book was the idea of Wayne Carpenter, writer, editor, publisher, and sometime voyager, who had trouble with his sextant one day at sea and made a mental note to get together a practical handbook on the subject. Not only did he conceive the project, find someone to write it, provide unrelenting encouragement and well-timed pressure, lay it out, type the manuscript several times over, advertise it, print it, haul it around, mail it off, and nurture it in every way, he sold out the entire first printing and moved on to other spheres and new challenges. His responsibility for this book was nearly total—he did everything but write it and illustrate it. His drive, energy, cheerfulness, and resourcefulness are phenomenal.

The drawings are the work of Nancy Bauer, artist of great skill and tact, whose notepaper, murals, and portraits are well known in Annapolis. Her drawings are exactly what I had in mind.

Many other people advised, taught, helped, and encouraged me. It's amazing how much can be learned by asking a few questions of the right people, and I have been extraordinarily lucky in finding them or already knowing them. Paul M. Anderton, Corinne Broekstra, Stephen Buraczewski, James Davis, Col. Warren P. Davis, G.D. Dunlap, Dr. Robert M. Fornili, OD, A.W. Fowler, Forrest W. Gibson, R.J. Holt, Frank J. Janicek Jr., Mitchell W. Kalloch, Cdr. John Luykx, Dr. Paul Millett (Cambridge University), Thomas Quinn, Hewitt Schlereth, Al Smith, Robert W. Selig, Jr., Robert Thompson, James Tindall, Dr. D.T. Whiteside (Cambridge University), Fred Gartzke, Eric Hiscock, MBE, H. Michael Newman.

Finally I would like to acknowledge my renewed appreciation of Daniel Boorstin's shrewd assessment, "I write to discover what I think."

Preface to the Second Edition

Sextant sales in the United States have declined since the First Edition appeared in 1986, and the national debt has increased enormously. While there is no apparent connection, it may be that the same sort of mental attitude is behind both trends—the lust for the easy answer and the quick fix. Why strain to balance the budget when more money can always be borrowed without difficulty or disapprobation—or mess about with an old-fashioned sextant and finical calculations when sleek electronic devices can spit out a position almost instantly without any tiresome figuring? Answers to both questions are evident to thoughtful people: In the end someone must pay for everything, and everything automatic sooner or later fails automatically, usually during or immediately before a crisis.

While electronic navigation has grown too comprehensive and convenient to be shunned completely by any but the most fanatical traditionalist, it continues to be vulnerable enough that it would be folly to abandon the practice of taking sights routinely. Merchant ship officers are required to do so. Off the Atlantic coast recently our radar, loran, and single sideband radio all were smoked in one brilliant instant by a lightning stroke merely near our vessel—not even a hit.

We who own and cherish sextants do not really care a nanoquark whether sales are up or down. One does not trade them in for new models, nor do they wear out from normal use. The longer you own one the better you like it and the more desirable it seems. Calculator navigation programs have eliminated the mathematical drudgery of tricky multiple interpolation in a maze of figures in tables. A sound sextant, and the ability to use it routinely, without dramatics, is like a balanced budget and money in the bank. It is a source of considerable security and satisfaction.

Contents

Introduction

CHAPTER 1

A SHORT HISTORY 17

Astrolabe
Latitude Hook
Quadrant
Cross Staff
Back Staff
Sextant

CHAPTER 2

**ELEMENTS OF THE SEXTANT
AND FOUR VITAL ADJUSTMENTS** 37

Perpendicularity
Frame And Index Mirror
Frame And Horizon Glass
Parallelism
Index Mirror And Horizon Glass
Telescope And Frame
Adjustment Procedures

CHAPTER 3

ATTACHMENTS AND ACCESSORIES 57

Astigmatizers
Double Star Prism
Wide View Horizon Glass
Davis Prism Level
Bubble Horizon Attachment
Neck Straps
Visual Aids

CHAPTER 4

CARE, MAINTENANCE AND REPAIR. **69**

Oiling And Cleaning
Immersion Bath Procedures
How To Hold
Where To Put Down
Mounted Sextant Boxes
Emergency Silvering Techniques
Applying Paint
Resilvering Horizon Glass

CHAPTER 5

HOW TO BUY A SEXTANT **83**

Where To Find Them
Prices
Five Ground Rules To Avoid Problems
Taking Test Sights
Interstellar Test Table
Discovering The Dropped Sextant
Reconditioning Costs

CHAPTER 6

SIGHTING TECHNIQUES **101**

Inversion For Location
Precalculating Azimuths
Rocking or Swinging An Arc
Sun's Upper Limb
Shades For Sun, Moon And Venus
Correcting For False Horizon
Rough Weather Sighting Techniques
Making A Monocle

CHAPTER 7

CORRECTING THE SIGHT **123**

The Thirteen Errors
Correcting For Moon, Venus And Mars
Correcting For Wave Height

CHAPTER 8

TIMING THE SIGHT ACCURATELY 137
The Hack Watch
Using A Tape Recorder
Counting Paces And Stopwatches
Attaching Stopwatch To Sextant

CHAPTER 9

SEARCHING FOR STARS 147
The Necessity For Preplanning
Finding Sunrise
Using The Rude Star Finder
Converting Local Time
Using *H.O. Pub. No. 249*
Making And Using A Wrist Board

CHAPTER 10

SEXTANTS OF TOMORROW. 157
Is The Sextant Doomed?
Coming Developments

APPENDIXES
A—Sextant Check Procedures 165
B—Sextant Manufacturers 166
C—Distributors and Dealers. 168
D—The Navigator's Basic Tool Kit 173
E—Making And Using An Artificial Horizon 175
F— Table of Interstellar Angles For Practice Sighting And
    Sextant Testing 176
G—Useful Addresses 178

BIBLIOGRAPHY. 181

INDEX 185

List Of Illustrations

CHAPTER 1

Figure 1-1 Latitude Hook
Figure 1-2 Arab Kamal
Figure 1-3 Modern Version of the Latitude Hook
Figure 1-4 Ancient Astrolabe
Figure 1-5 Quadrant
Figure 1-6 Cross-staff
Figure 1-7 Cross-staff With Two Arcs
Figure 1-8 Holt Sighting With Backstaff
Figure 1-9 Nocturnal
Figure 1-10 Newton's Double Reflecting Instrument
Figure 1-11 Hadley's First Instrument
Figure 1-12 Godfrey's Instrument
Figure 1-13 Hadley's Second Instrument
Figure 1-14 Bauer Sighting With Hadley Instrument
Figure 1-15 Double Reflectivity Diagrams

CHAPTER 2

Figure 2-1 Sextant Parts Identified
Figure 2-2 Sextant Disassembled
Figure 2-3 Arc, Micrometer Drum and Vernier Scales
Figure 2-4 Cassens and Plath Sextant With Fulvew Horizon Glass
Figure 2-5 Left Hand Sextant
Figure 2-6 Perpendicularity of Index Mirror
Figure 2-7 Alternative Index Mirror Check
Figure 2-8 Side Error Check Diagram
Figure 2-9 Side Error Check With Star

CHAPTER 3

Figure 3-1 Astigmatizer View of Star
Figure 3-2 Davis Prism Level
Figure 3-3 Bubble Sextant Field of View
Figure 3-4 Neck Strap

CHAPTER 4

Figure 4-1 Sextant Hook
Figure 4-2 Sextant Box Security
Figure 4-3 Horizon Glass Unmasking

CHAPTER 5

Figure 5-1 C. Plath Navistar Classic
Figure 5-2 Cassens & Plath With Bubble Attachment
Figure 5-3 Tamaya Jupiter
Figure 5-4 7/8 Tamaya Yacht Sextant
Figure 5-5 Weems And Plath Sextant With Star Scope

CHAPTER 6

Figure 6-1 Sextant Inverted For Acquisition
Figure 6-2 German Gyro Sextant
Figure 6-3 Slant Distance Diagram
Figure 6-4 Rendition Of View While Rocking Sextant
Figure 6-5 Shade Glasses On Zeiss Freiberger Sextant
Figure 6-6 Dutch Observator With Integrated Shades
Figure 6-7 False Horizon Created By Moon
Figure 6-8 Star Split By Horizon
Figure 6-9 Navigator's Monocle

CHAPTER 7

Figure 7-1 Cassens and Plath Sextant Certificate
Figure 7-2 British NPL Certificate
Figure 7-3 Tamaya Inspection Certificate
Figure 7-4 Index Error Readings
Figure 7-5 *The Nautical Almanac* Additional Correction
Table, A4
Figure 7-6 Altitude Correction Guide
Figure 7-7 Correction For Wave Height

CHAPTER 8

Figure 8-1 WWV and WWVH Time Signal Format
Figure 8-2 Time Tick Tape Recorder
Figure 8-3 Split Second Hand Stopwatch
Figure 8-4 Cassens and Plath With Timex Lap Timer
Attached

Figure 8-5   Heuer Microsplit Lap Timer

CHAPTER 9

Figure 9-1   Rude Star Finder Kit
Figure 9-2   Star Finder Base Plate And Template
Figure 9-3   Page From *HO 249*
Figure 9-4   Precalculated Star Location Data on Wrist Board

CHAPTER 10

Figure 10-1 Cel Nav Electronic Sextant
Figure 10-2 C.Plath Electronic Sextant

Introduction

Ownership of sextants is widespread, but proficiency in their use and adjustment is not. Some owners are, at heart, afraid of their sextants in the same sense that I would be fearful of handling the *Dead Sea Scrolls* lest they crumble to dust in my clumsy grasp. The instrument is considered by some to be so delicate that merely taking it out of its box is audacious and the idea of actually using it at sea like any other tool of the mariner's trade seems almost callous.

There is no doubt that a sextant is fragile and cannot be banged carelessly around and remain capable of precise measurement. It is a basic theme of this book, however, that sextants are remarkably tough and durable machines that, with reasonable care, will not only outlast their present owners, but several successive generations of their seafaring descendants as well, even with daily use. The number of serviceable instruments around today that were manufactured in the eighteenth century is testimony enough.

The first part of this book is devoted to the antecedents and history of the sextant, while the second is devoted to practical matters. The instrument's name is from the Latin *sextans*, meaning one sixth of a circle. We call it this in spite of the fact that ever since the early eighteenth century optical manipulations have enabled instruments to measure twice as many degrees as the physical length of their arcs. It is all done with mirrors. A sextant can measure 120 degrees or more and technically should be called a *tritant*. The instrument Columbus used actually was a *quadrant*—a quarter of a circle both in measuring capability and length of arc. It turned out to be about as useful as the Arab translator he took along to speak to the emperor of

Japan. The first modern sextant-type instruments were octants in that they had arcs of 45 degrees and could measure 90 degrees.

Certain parts of the sextant are precisely adjusted and delicate. Generations of sailors have been taught not to mess about with the mirrors, for example, except as an absolute last resort. For many years, I was too timid to clean the mirrors of a sextant for fear of putting them out of alignment. I must have missed many a star sight for being unable to see the reflection in the scum accumulated on the mirrors. Not only are the mirrors sturdily mounted and readily cleaned without distresssing them in the least, but they are easy to adjust. The trick is not only to know how, but when you should and when you should not. This book will tell you that, as well as how to use the instrument to its full potential, maintain, repair and stow it. It may also be found helpful in deciding what sextant to select in the first place.

Cmdr. Bruce A. Bauer, USN (Ret.)
Annapolis, Maryland
March 7, 1986

CHAPTER 1

A SHORT HISTORY

Perhaps the earliest and certainly the simplest of the ancient antecedents of the measuring instrument we today call the sextant was the latitude hook of the Polynesians. It too was used to measure the distance between a celestial body and the horizon. Those primitive navigators carried a set of hooks to be used during different legs of a voyage. One could sail perfectly well today from Lisbon to the Azores to the Delaware Bay capes with but one latitude hook, for those places are at the same latitude. Or, altering course for Boston from the Azores, one would take up a different, somewhat longer latitude hook to match the higher altitude of Polaris in the latitude of that destination. **Figure 1-1** shows a hook aligned.

A working latitude hook can be made from a piece of stiff wire. Bend a small hook or loop in one end of the wire and then sight Polaris, the pole-marking star, through the loop with the tail of the wire hanging vertically. Where the wire is crossed by the horizon a mark or an L-shaped bend is made and you will have recorded the latitude of that position, which might be a vessel's point of departure. Any time later, having traveled east or west, if the instrument will just fit between the polestar and the horizon, the observer is at the same latitude. If it will not, he must move north or south to correct and make the hook just fit the gap.

Ancient Arab navigators used an instrument working on the same principle but of somewhat greater refinement called the *kamal*. The word means guide in Arabic. A flat rectangular piece of wood was cut to a size to fit between the horizon and the North Star when seen from home port. To keep the wood at a uniform arm's length from the eye, a string was attached to a hole in the center of the board. A knot

17

in the string could be held in the teeth while the string was stretched taut. Other knots could be added for the latitudes of various destinations so that the kamal was more versatile than the latitude hook and more accurate as well. It is still in use on dhows navigating off east Africa.

A modern version of these two instruments is used by Col. Warren Davis when sailing up the U.S. East Coast making yacht deliveries. It is a right angle made of rosewood with the vertical part notched for various latitudes—St. Augustine, Charleston, Morehead City, Cape Henry, Sandy Hook. The end of the horizontal base of the right angle is held against the nose, aligning the sighting eye and the North Star sighted through the appropriate latitude notch. Fairly accurate approximations of the latitude can be made—particularly if allowance is made for the day, time and corresponding displacement of the star from the pole. See **Figure 1-3**.

The simplicity, economy and durability of these instruments were good, but they were awkward to use because they required looking two places at once—the body and the horizon. At least they could be used at sea—an attribute sometimes sadly lacking in a later development—the *astrolabe*. An elegant instrument, there were some models that looked more like jewelry or artful decorations than practical measuring devices. Suspended pendulum fashion by either the navigator or his assistant (preferably a tall chap), the astrolabe measured the degrees down the celestial sphere from the zenith directly over the observer's head to the body. Subtracted from 90 degrees, this value, called the *zenith distance*, yielded the altitude of the body. It was also the approximate latitude, if the pole star were the body sighted. The trouble arose from the requirement that the astrolabe had to be motionless to establish the vertical, not to mention lining up the sight vanes. It worked well for the desert traveler who could simply get down off his camel to attain a stable platform, but the ocean navigator had to interrupt his voyage, ferry equipment ashore and take sights from there. Many thought that was too much trouble and just pressed on with fingers crossed. See **Figure 1-4**.

Aboard ship, a device called a *quadrant* could tolerate a small amount of motion. It worked on the same principle as the astrolabe— measuring the angle between the vertical and the line of sight to the body. It was a quarter of a disk with a scale along the curved arc, a plumb bob suspended from a pivot point, and sighting vanes on the top edge. Again, assistance was required since the person sighting

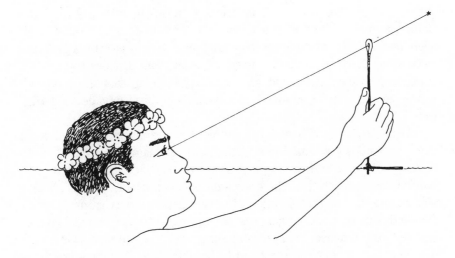

Fig. 1-1 Latitude Hook

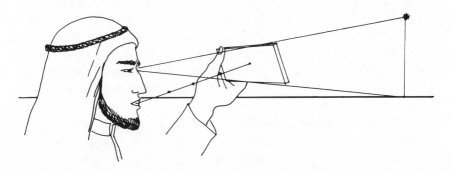

Fig. 1-2 Kamal

Fig. 1-1 A latitude hook, made of split bamboo and shell, was aligned with the horizon and pole star to show when the observer was at desired latitude.

Fig. 1-2 The kamal, meaning "guide", was a wood block held at a predetermined distance from the eye with the help of a knotted string. When it just fit between the pole star and horizon, it showed that the observer was at the same latitude as that of the destination.

could not simultaneously see the scale. Imagine the teamwork required on a rolling deck to keep the sights on the sun and read the value between swings of the plum bob. This instrument must have been used mostly while lying to or anchored. The quadrant's greatest virtue may have been that it inspired exasperated users to demand something better. As crude as it was, though, it was the most advanced instrument Columbus had aboard on his first voyage. See **Figure 1-5**.

The horizon is a handy reference at sea and can be seen most days. To some, it seemed more practical to measure up from the visible edge of the world, in degrees of arc to the body, to obtain its altitude. To accomplish this, the *cross-staff* was contrived, very likely by a ship's carpenter at the direction of some desperate navigator. It was a simple device of wood with a crossbar arranged so as to be slid along a squared staff that was marked off in degrees. This, at last, was a one-man instrument, but the user had to be able to look in two directions at once and with the same eye. The butt end of the staff was placed on the cheek against the side of the nose beside the eye and the crossbar moved outward until it just fit in the space between horizon and the body like the wood plate of the kamal. The scale on the staff converted inches, which were being measured, to degrees of altitude by trigonometric function (tangent of half the altitude equals half the crossbar length divided by eye-to-crossbar distance). There were interchangeable crossbars for various ranges of altitudes—longer bars for higher altitudes—and different scales for each on different side faces of the staff. See **Figure 1-6**.

The great advantage of this instrument was its flexibility in that it could measure varying altitudes, which the earlier instruments could not. A proficient operator could get relatively good sights. In fact it still is potentially useful in emergencies. In the chapter on lifeboat navigation in *American Practical Navigator*, by Nathaniel Bowditch, directions are given for the makeshift construction and use of the cross-staff for determining the altitude of bodies and for distance away from geographical features of known dimensions as well.

The cross-staff was too awkward to endure unmodified. The worst thing about it was that persisting vexation—the requirement to look two places at once. This is a hard task; the eye can concentrate sharply on only one spot at a time. Ocular parallax could cause errors of as much as one and a half degrees—equal to as much as 90 miles. Cross-staff sights were only approximations by modern standards,

Fig. 1-3 Col. Warren Davis, of Annapolis, Maryland, demonstrates his modern version of a latitude hook. He has used it successfully on yacht deliveries up and down the East Coast.

but then navigation itself was an imprecise art until techniques were developed and refined. One such practical development was the deliberate introduction of error to eliminate later doubt.

In running down a latitude—that is sailing east or west while maintaining the same latitude line—the navigator knew he probably was going to miss the precise destination. The question was, in which direction, north or south? Which way to turn when arriving at that uniform coastline with no distinguishing landmarks was the problem, so the navigator deliberately steered a few degrees north or south of the course in order to be fairly sure of which way to turn upon making landfall.

The next development, a giant step, was the *backstaff*, invented in the comparatively recent year of 1590, by a man named John Davis, a seeker of the Northwest Passage to the Pacific, for whom Davis Strait in Canada was named. This instrument permitted the navigator to match the shadow of the sun to the line of the horizon. To do so, he stood· with the sun behind and over one shoulder and sighted the horizon through a movable peep hole attached to a large arc; the shadow of the sun was admitted through a slit in another movable vane on the smaller arc close to the front of the instrument. One could adjust both vanes to achieve coincidence. With a third variable of the motion of the ship added, the navigator must have had a lively time. The sum of the readings on the two arcs equaled the altitude of the sun. In order to take sights of other bodies, which cast no shadows, a mirror was added, but that came later. See **Figures 1-7** and **1-8**.

To get to this stage of coincident viewing of body and horizon had taken centuries. John Davis, who was a working mariner and not a theoretician, was a remarkable combination of sailor and scientist. His book, *The Seaman's Secrets*, published in 1595, was . . *.for sailors, not for scholars on shore.* It was the first of a kind of nautical manual intended to guide the user through a sequence of actions to produce a result—a sort of navigational cookbook. Previous books on the subject

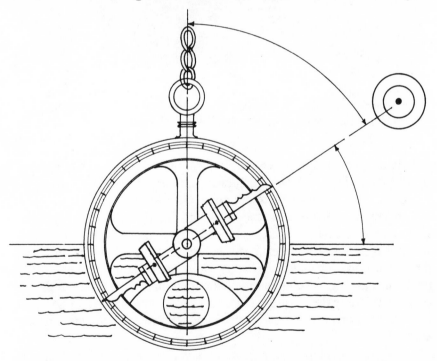

Fig. 1-4 The astrolabe, or "star taker", gave the altitude of a body by measuring its complement from the zenith down to the line of sight.

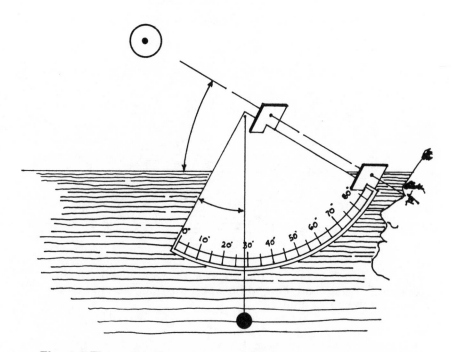

Fig. 1-5 The quadrant, accurate only when motionless, was of limited use at sea.

tended to inspire something more like scholarly contemplation of the intricacies of the heavens, while Davis wrote simply to help seamen find the way out and back.

Whether the instrument used was a backstaff or a cross-staff, when measuring the altitude of Polaris for latitude, a correction for its varying distance away from true north was necessary. Before the corrections were tabulated in *The Nautical Almanac*, they were determined with the *nocturnal*. See **Figure 1-9**.

This early metal instrument was shaped something like a lady's hand mirror and usually elaborately decorated. It had a hole in its center through which the pole star was sighted. A pointer like a long clock hand was aligned with the nearby star Kochab, which circles the pole. The nocturnal's scale yielded, in minutes of arc, the correction to be applied to the sighted altitude of Polaris. Polaris' displacement from the pole could result in a 50-mile error if not taken into consideration. Because it was easy, this method of finding latitude was popular. Recently, however, latitude by Polaris has declined in importance.

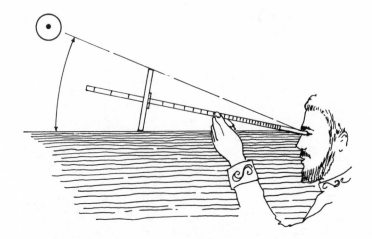

Fig. 1-6 The crossbar of the cross-staff was moved out the staff until it just filled the gap between the horizon and sun to measure the sun's altitude.

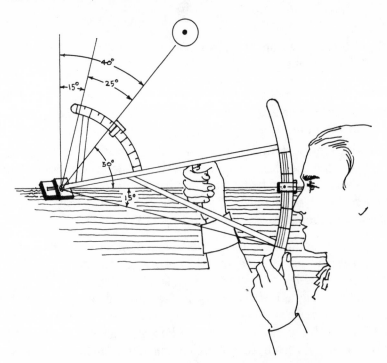

Fig. 1-7 With this model of the backstaff, the altitude of the sun was obtained by combining the readings from the two arcs of the instrument.

The simplicity of the calculation does not change the fact that Polaris is a dim star (magnitude 2.1) and hard to measure against anything less than a knife-edge horizon under the best conditions.

Fig. 1-8 R.J. Holt, director of the Chesapeake Bay Museum at Saint Michaels, Maryland, sights with an authentic Davis backstaff, which he crafted from Smithsonian Institution plans.

Nocturnals, which faded from use in the late fifteenth century, also could be used to derive time from the celestial clock which pivots around Polaris, after the month and day of the year were set on scales at its perimeter. This star clock could be read only to a rough approximation of the time, not good enough for modern navigation.

The backstaff had been in use for more than a hundred years when there apparently occurred one of those coincidences, which happen from time to time in the progress of science. Three men working independently arrived at essentially the same solution to the problem of how to make an instrument suitable for measuring angles between celestial bodies and the horizon at sea.

In 1699, Sir Isaac Newton (1642-1727) conceived the principle of measuring angles by double reflection and made an instrument that would do so. At the time, he was the most prominent member of the Royal Society of London for Improving Natural Knowledge, known then, as now, simply as The Royal Society. Chartered first in 1662, this group of the world's most distinguished scientists met regularly to promote scientific discussion and encourage further investigation.

It also advised the government, from which it received a subsidy. Another purpose of The Royal Society was to act as a forum for establishing priority, that is individual credit, for discovery or invention. The Royal Society was closely involved in a variety of scientific inquiries of which the development of a means to determine longitude at sea was one of the most prominent.

On 9 August 1699, the celebrated Newton, the Einstein of his age, personally appeared before The Royal Society to reveal a new principle of optics to this group whose purpose was to receive, evaluate and disseminate scientific knowledge. He read and submitted a paper on an instrument made on the principle. From the Journal Book of The Royal Society (minutes of the meeting), set down in a bold and flowing script:

> Mr. Newton shewed a new instrument contrived by him for observing the moon and starrs, the longitude at Sea, being the old instrument mended of some faults with which notwithstanding Mr. Hally (sic) had found the longitude better than the Seamen by other methods.

The minutes were referring to Edmund Halley (1656 – 1742), later secretary of The Royal Society, who applied Newton's concepts to predict the comet that bears his name. In 1699 Halley had recently returned from a voyage to Brazil, during which he evaluated an instrument designed by Dr. Robert Hooke, a rival of Newton. Hooke's instrument used a mirror, but did not incorporate the double reflecting principle. It did not work very well.

This Royal Society Journal Book entry constituted a clear record of what in current patent office terminology would be called disclosure of a new invention. That someone was listening attentively is demonstrated by the insertion of a comment in the minutes of the next meeting—25 October 1699—by Dr. Hooke that:

> ...the instrument mentioned last meeting was of his own [Hooke's] invention before the year 1665 and that the use and fabric of it was declared in the History of the Royal Society.

He disputed the originality of Newton's concept or model or both. We cannot tell which.

John Hadley (1682-1744), the London mathematician and instrument maker who is generally credited with being first to produce a double reflecting instrument, was nineteen when Newton made his

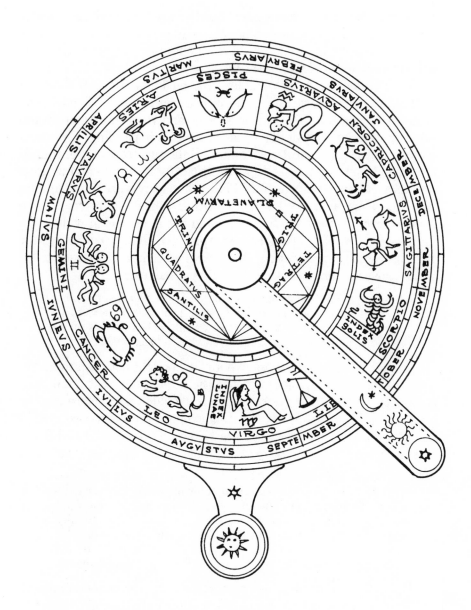

Fig. 1-9 The nocturnal, often elaborately decorated, was used to measure the degrees of arc between the pole star and true north at any particular time.

disclosure. Thirty-two years later, Hadley, who had become a leader in his field and a vice president in the Royal Society himself, presented a paper on his version of the double reflecting instrument and two models to demonstrate it.

It is widely held that Hadley had no knowledge of Newton's application of the same principle even though the two lived in the same city, belonged to the same society and were interested in the same specialized subject. The reason for this is puzzling. The usual explanation is that secretary Edmund Halley somehow absentmindedly misplaced the plans and specifications for Newton's invention—and they did not reappear until 1742, when they were found among the Halley papers after his death. Other considerations may have some bearing on why Newton's idea did not at once gain wide exposure. When The Royal Society was organized, Henry Oldenburg, the first secretary, appreciating the "growing concern" with priority, made provision for protecting the rights of members:

> When any Fellow have any philosophical notion or invention not
> yet made out, and desired the same, sealed in a box, to be
> deposited with one of the secrataries till perfected, this might
> be allowed for better securing inventions to their authors.

With this arrangement an inventor who might not wish to implement a device could file it away with the Society for what could later constitute a prior claim. Newton died some four years before Hadley brought out his instruments. Newton had been president—some said dictator—of The Royal Society from 1703 until his death. Daniel Boorstin, in his book, *The Discoverers*, continues describing the great man:

> As his prestige grew so did his dyspepsia, his unwillingness to
> give credit to others or share credit for his great discoveries. To
> assert his primacy in every branch of science that he touched he
> marshalled his full powers over what had been called the first
> scientific establishment in the modern world ... he never toler-
> ated any sign of "levity or indecorum," and actually ejected Fel-
> lows from the meetings (*of The Royal Society*) for misbehavior.

It is well to remember also that in the early seventeen hundreds, Britain was expanding, exploring, competing and contending with the other powers. British seamen were pushing out the boundaries of the Empire and as yet there was no accurate means of marking the east/west position of a place.

In 1714, a petition from a group of sea captains and merchants was received by Parliament which expressed the seriousness of the deficiency:

Fig. 1-10 Newton's 1699 double reflecting instrument.

> That the discovery of longitude is of such consequence to Great
> Britain, for the Safety of the Navy, for Merchant Ships, as well
> as for improvement of trade, that for want thereof many ships
> have been retarded in their Voyages and many lost . . ."

In 1715, Parliament offered a cash prize for development of a
practical method of determining longitude at sea—the amount
awarded to depend on the accuracy achieved— £10,000 if within one
degree, up to £20,000 if within one half degree. There were only two
methods for determining longitude known to be theoretically feasible
at that time—lunar distances and portable timekeepers. Newton's
1699 design may well have been put on hold awaiting the advent of a
method of putting it to use that would both be practical and enable
participation in the prize money. There also may have been an
element of concern for national security in the long sequestration of
Newton's design.

The sextant could be considered a kind of a secret weapon capable
of conferring a real advantage in exploration, commerce and military
prowess. It was almost certain to be a vital element in solving the
problem of finding longitude—and so it eventually was.

Thomas Godfrey (1704-1749) was a Philadelphian—a colonial. He was a mathematician, glazier and optician, and a live-in tenant in Benjamin Franklin's Market Street home. It was Godfrey who did moon and tide calculations for some of Franklin's early almanacs.

He was a member of Franklin's scientific and philosophical group, *The Junto*, for a while. Here is Franklin on Godfrey:

> "... a self taught mathematician, great in his way, and afterward inventor of what is now called Hadley's quadrant. But he knew little out of his way, and was not a pleasing companion; as, like most great mathematicians I have met with, he expected universal precision in everything said, and was for ever denying or distinguishing upon trifles, to the disturbance of all conversation. He soon left us."

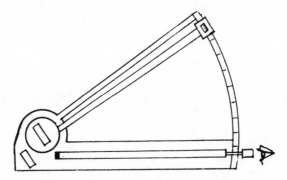

Fig.1-11 Hadley's first double reflecting instrument was inverted compared to Newton's, but it used the same concept of a fixed telescope and a mirror rotated by an index arm.

Godfrey, who sounds like the archetypical inventor, did have some friends, among them sea captains who explained the problems in taking sights at sea. In pursuit of the unclaimed reward for longitude, Godfrey first modified a wooden quadrant and then designed and built from scratch a double reflecting instrument which he sent to sea for testing in 1730. Results were good enough to encourage him to submit the design to The Royal Society the same year. He received no acknowledgement, as far as we know.

Then, in 1731, The Royal Society published a description of Hadley's instruments, never mentioning Godfrey. John Logan, a prominent Philadelphia intellectual, companion of Penn, friend of Franklin and Godfrey, and sponsor of The Junto, was a respected correspondent of The Royal Society. He collected documentation of the chronology of

Godfrey's invention and pressed Godfrey's case, writing to Halley and other friends in the society. The society decided finally that this was a case of simultaneous, independent invention and that both men deserved recognition. The prize money was divided equally.

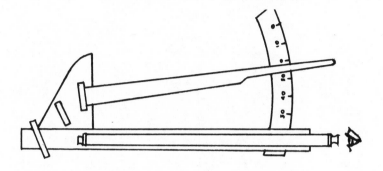

Fig. 1-12 Godfrey's double reflecting instrument.

Both men had designed instruments using the identical double reflecting principle first detailed thirty years earlier by Newton. This is a coincidence, but not one of the first magnitude. After all, both Hadley and Godfrey were mathematicians and makers of optical instruments and the principle they incorporated was a logical step in the development of the backstaff once a mirror had been attached. What is probably more remarkable is that Newton, with his mania for getting the credit for his ideas, allowed this one to lie dormant for so long.

The last years of Newton's life were devoted to rationalizing the apparent contradiction between the teaching of Christianity, which he cherished, and the implications of his own scientific principles. He was, according to Dr. D.T. Whiteside, distinguished Cambridge professor and Newton authority, a "good inland towns man . . . (who) never saw the sea . . . never had any wish to."

Regardless of who deserves the distinction of inventing the first double reflecting sextant, there is no controversy over who carried the principle on to produce practical improved instruments—John Hadley.

He went into production with instruments on the lines of the second of the two instruments he had showed the society in 1731—the instrument with a horizon glass which was well suited for matching the celestial body to the horizon. See **Figure 1-13**.

Hadley's second instrument had an arc of one-eighth of a circle, making it an octant and enabling it to measure up to 90 degrees because of the double reflectivity bonus. It also differed from earlier instruments in that the arc was situated at the bottom of the frame. It looked like a tall, narrow, modern sextant.

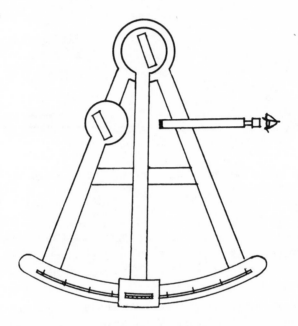

Fig. 1-13 Hadley's improved double reflecting instrument. The arc moved down to the bottom where it has been ever since.

What is this principle of double reflectivity? It depends on the observable fact that in bouncing off the surface of a mirror, a ray of light departs at the same angle at which it arrived. The angle of incidence equals the angle of reflection. See **Figure 1-15A**.

If a ray of light is bounced in sequence off two mirrors that are precisely parallel, the same in-out equality will be maintained. This is the condition achieved by a sextant in true adjustment set on zero degrees. See **Figure 1-15B**.

The sextant is, in essence, a machine for varying the angle between two mirrors by precisely measurable numbers of degrees to

Fig. 1-14 Author Bruce Bauer takes a practice sight with a replica of Hadley's second instrument. Photo by R.J. Holt.

utilize the phenomenon that the angle of the departing light ray will have been changed by double the angle between the mirrors. If a body is 70 degrees above the horizon, the parallel mirrors of the sextant must be displaced by 35 degrees in order to send its reflected light rays straight through the optics. See **Figure 1-15C**.

The scale on the limb of the sextant arc is calibrated to read 70 degrees for the 35 degrees of motion displacing the index mirror at the upper end of the arm. An important reason for the requirement for extraordinary precision in the manufacture of sextants is that any error that may be generated by the displacement of the mirror also will automatically be doubled in magnitude.

Since the birth of the modern sextant in 1731, there have been periodic improvements, and the cumulative result has been today's instrument, marvelously refined and sophisticated compared to the Hadley and Godfrey instruments. But there have been no radical departures or new principles introduced since. In 1733, Hadley himself introduced a significant refinement by attaching a spirit level to his sextant so the horizontal plane could be determined even when the horizon could not be seen. In the 20th century, sextant makers such as C. Plath, Cassens & Plath, Tamaya, and Davis have each brought out versions of a wide-view horizon glass, the value of which is not entirely beyond debate. These developments, along with tangent screws and enlarged mirrors, are but refinements of the two early instruments, which could still serve the modern navigator adequately—in contrast, for example, to the relationship between an abacus and today's electronic computers.

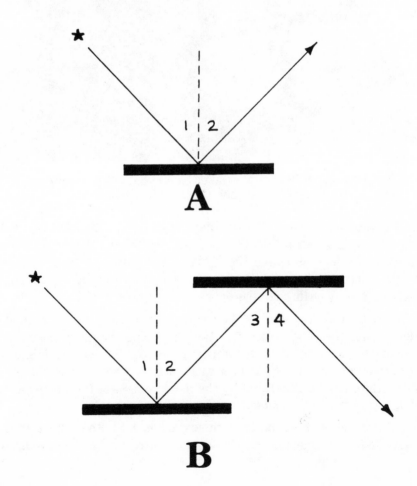

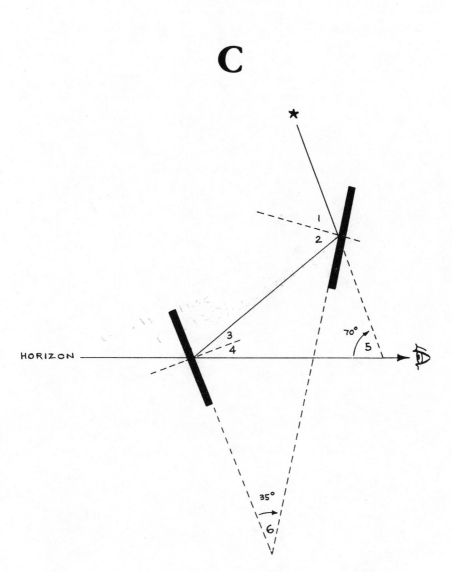

Fig. 1-15 Double reflectivity depends upon the principle that the angle of incidence of a ray of light equals its angle of reflection (angles 1 and 2 in View A). A twice-reflected image will maintain that equality as long as the mirrors are parallel (angles 1, 2, 3 and 4 in View B). When one of the mirrors is rotated away from the parallel, the total change in the angle of reflection from both mirrors is twice the amount of rotation (angles 5 and 6 in View C).

CHAPTER 2

ELEMENTS OF THE SEXTANT
AND FOUR VITAL ADJUSTMENTS

A—Frame: The frame is the platform, base plate or main trunk to which all other parts are attached and depend for stability. Its chief characteristic must be its rigidity. It could be made of other staunch, firm materials but generally is made of brass, or brass alloy, for its nobility, inertness and resistance to corrosion. Modern frames sometimes are made of aluminum alloy for lightness or plastic for cheapness, lightness and resistance to corrosion. Frames of early instruments were made of ebony or rosewood. The frame should not expand and contract significantly with temperature nor warp if left in direct sunlight so that one side is hot and the other cool. Plastic sextant frames are vulnerable in this regard and must be given special handling.

The classic frames are of cast metal which are later machined to the most precise specifications. Frames can be solid, as is customary in plastic instruments where lightness is inherent, or webbed—ventilated, in metal instruments, like aircraft construction—to reduce weight. Artistic frame design is a point of pride with leading sextant manufacturers.

B—Limb: The limb is an integral part of the frame and forms its curved, reinforced lower edge. Though it looks like a solid bar when viewed from the left side, it is usually L-shaped in cross section.

Along its bottom edge are precision machined gear teeth or diagonal slots, one for each degree which the instrument is de-

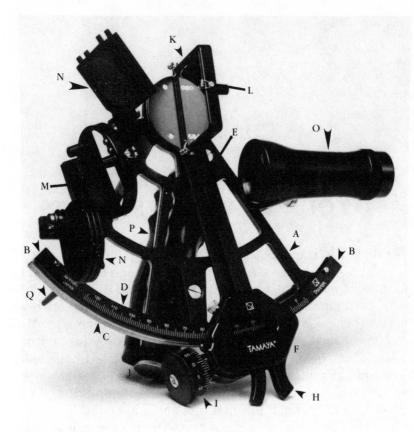

Fig. 2-1 Parts Of A Sextant Identified. Photo courtesy of Baker, Lyman & Co.

A - Frame	J - Battery Compartment
B - Limb	K - Index Mirror
C - Gear Rack	L - Adjusting Screw
D - Arc	M - Horizon Glass
E - Index Arm	N - Shade Glasses
F - Index Mark	O - Telescope
H - Clamp Release	P - Handle
I - Micrometer Drum	Q - Leg

Fig. 2-2 Tamaya disassembled to show main subassemblies, tapered pivot pin (E-1), and tangent worm screw (G). Photo by Scofield & Wolfe.

A - Frame
A-1 - Pivot Bearing
B - Limb
C - Gear Rack
D - Arc
E - Index Arm
E-1 - Pivot
G - Tangent Worm Screw
H - Clamp Release
I - Micrometer Drum

M - Horizon Glass
N - Shade Glasses
O - Telescope
O-1 - Circular Collar
O-2 - Telescope Collar Bracket
P - Handle
Q - Legs
S - Index mirror

signed to measure. Collectively, they are called the gear rack, (C). There are more teeth than degrees because the limb is machined from end to end, extending beyond the scale of the instrument. The trueness of the shape of the limb and the accuracy of the machining are the basic qualities that govern the value of the sextant.

The length of the limb usually is the sixth part of a whole circle, though it may be more or less.

D—Arc: The arc is the scale of degrees attached to the left face of the limb, or cut right into the body of the limb itself. Its lines mark degrees of arc and are precisely aligned with the teeth on the bottom edge of the limb. The modern sextant scale commonly reads up to about 120 degrees and down to about minus five degrees. This wide range is of value chiefly for measuring horizontal angles between landmarks, as in surveying, rather than in celestial measurements.

The legibility of the arc of a sextant is important. If the engraving is shallow and the contrast between the marks and the background not great, misreadings will be more frequent. Some sextants have painted arcs with white lines on black, or the reverse. While I prefer a classic, plain brass, engraved arc that is integral to the limb, I admit that toward the end of twilight it is considerably easier to read one of the new high contrast arcs.

E—Index Arm: The index arm is the most prominent moving part of the sextant. Both its *pivot* (**E-1**) and drive arrangement along the limb are critical for precision. The arm is pivoted at the upper corner of the triangular shaped frame at the exact center of the curve of the limb. If the pivot is off center, the effect is like that of a wheel spinning on a bent axle—eccentric and unsatisfactory. The index arm extends beyond the limb forming a plate that has an aperture with an *index mark* (**F**), on its lower edge. This mark is matched to the scale of the arc for a reading.

G—Tangent Worm Screw: On the lower end of the index arm is the tangent worm screw arranged so as to move the index arm along the gear rack of the limb as it is turned. The tangent screw is held engaged against the limb gear teeth by spring pressure and can be held back away from the gears disengaged by means of a *clamp release* (**H**). When disengaged, the arm can be moved freely and quickly along the arc for initial general alignment.

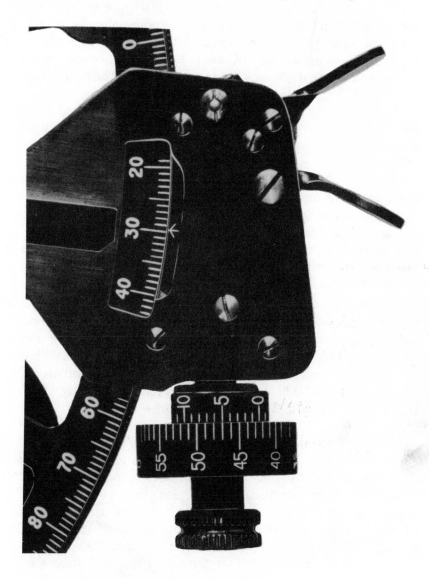

Fig. 2-3 Micrometer drum and adjacent vernier scale graduated in tenths of minutes of arc. The reading is 29, from the main scale on the arc, 42 minutes from the micrometer drum adjacent to the 0, and .5 minute from the vernier—29°42′.5. Photo courtesy of Defense Mapping Agency.

I—Micrometer Drum:

At the end of the tangent screw shaft is the micrometer drum which turns one full revolution for each degree of the arc traversed. The micrometer drum is graduated in minutes of arc. On the right of the micrometer drum is a short vernier scale calibrated as to read in seconds or in tenths of minutes. Because *The Nautical Almanac* deals in degrees, minutes, and tenths, the latter is more convenient, but mental conversion is easy.

The micrometer drum is a major development that gives a great advantage over the old, long vernier-against-the-arc scale arrangement. The possibility of an erroneous reading is greatly reduced.

K—Index Mirror: The index mirror is at the top end of the index arm and rotates with the arm as it is moved. This varies the angle of bounce of the incoming light by amounts shown on the arc scale. The index mirror has an *adjusting screw* (**L**) so that its perpendicularity to the arc can be adjusted. The center of rotation of the mirror viewed edgewise is directly aligned with the pivot point of the index arm.

M—Horizon Glass: The horizon glass is mounted on the front edge of and is perpendicular to the frame. It is silvered on its right half to make it a mirror (on the side toward the frame) while the left half remains clear glass in order to permit a view of the horizon. Some newer horizon glasses are not silvered but quartz-coated to be both reflective and transparent over the entire surface. **Figure 2-4** shows how large an opening this provides.

N—Shade Glasses:

Both the index mirror and the horizon glass are provided with shade glasses to enable the navigator to look directly at the sun without injury or discomfort. These glasses come in varying degrees of density for use in varying light conditions. The glass of which they are made must be optically true—that is, with faces parallel—or else error will be introduced. They are expensive—a replacement set of four for a C. Plath instrument costs about $275.

O—Telescope: A telescope is rigidly attached to the rear edge of the frame so that its line of sight is parallel to the frame and passes through the center of the horizon glass.

Two different scopes are often provided for different conditions and sighting targets. A bell-shaped scope of three or four power, called a star scope, is used at twilight when too much magnification of a star might obscure an indistinct horizon.

The most common higher powered scope is a 6x30 monocular which is used for sun sights and in conditions when discerning the horizon is not a problem. Its field of vision is smaller than that of the star scope and the body appears much larger, both directly and in proportion to the field of view.

A lensless tube may be used to direct the line of sight without any magnification. Some instruments can be used with no sighting device at all; the line of sight is simply directed through the circular collar (O-1) into which the scope ordinarily is fitted. The accuracy of the sight is not impaired as long as alignment is true and one's eyesight is keen enough to see the body distinctly. A disk with a hole in the center fitted into the scope collar will facilitate true alignment. The disc can be made of machined brass or cardboard. O-2 in **Figure 2-2** is the bracket used to attach the telescope collar to the frame. If it is adjustable so that the scope can be moved toward or away from the frame, its proper name is the *rising piece*. Its function is to vary the brightness of a star relative to the distinctness of the horizon. The horizon, when dim, cannot be seen as well if the star being measured is too bright.

P—The Handle: This usually is made of high quality molded plastic, although one or two companies still offer the classic carved wood. Handles usually are hollow with space for batteries for illumination of the arc and micrometer drum. Some are molded to fit the hand more comfortably and may be attached to the frame at different angles so that wrist twist is minimized. Some newer models may have adjustable handles. The handle also may be designed to accommodate a timing watch.

Until recently, every marine sextant I had ever seen had the handle on the right side of the frame. As far as I am concerned, this arrangement is most convenient for left handed navigators since the rest of us must put the instrument down or at least shift it to the other hand in order to record each sight. This is

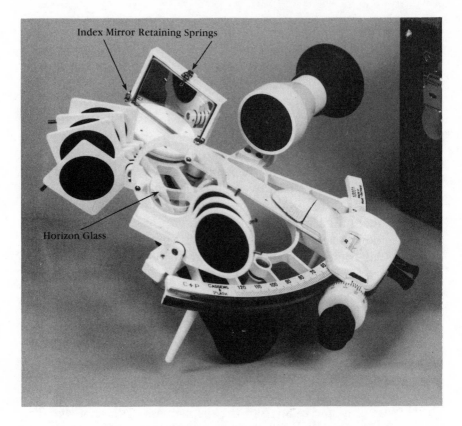

Fig. 2-4 The Cassens & Plath Admiral Fulvew horizon glass is both reflective and transparent. It has no vertical division in the center. Similar horizon glasses are optional on many brands of instruments. Photo courtesy of Baker, Lyman & Co.

awkward. It's like a right handed ball player wearing a glove on his right hand which he has to take off in order to throw the ball.

Recently, I came across a left handed marine sextant made by Brandis and Son, of Brooklyn, New York, in an antique instrument shop at Annapolis, Maryland. **Figure 2-5** shows store owner and sextant authority G.D. Dunlap, sighting through the instrument.

Dunlap said he has never seen another one in his many years of experience. Notice there is nothing awkward about the handle being on the left side. Either eye can be used and the right hand is free to make notes.

The instrument even has an ivory plate on the handle on which to write. I lifted the instrument and found it comfortable. I have been unable to learn why marine sextants are all made with the handle on what is, for most of us, the wrong side. Some aircraft bubble sextants do have handles on the left side.

Q—Legs: The legs are another feature that may be on the wrong side of the instrument. They are on the same side as the handle, which means that it is difficult to avoid picking up the instrument by some other part first in order to get at the handle. If the legs were on the mirror side of the instrument it could be put down with the vulnerable mirrors arched over by the frame and protected. Ah, tradition!

Four Vital Adjustments

Four cross-checks should be routine practice before each round of sights. Together they take but a minute or so and ensure the instrument is really ready to be used.

They are particularly desirable if someone else has used the instrument last. These checks and adjustments are:

1. **Perpendicularity—Frame and Index Mirror**
2. **Perpendicularity—Frame and Horizon Glass**
3. **Index Error**
4. **Parallelism**

The first two checks for perpendicularity determine whether the two reflecting surfaces are at right angles to the main platform of the instrument.

Fig. 2-5 G.D. Dunlap, navigation instrument authority and author of *Successful Celestial Navigation with H.O. 229*, sights with a rare left-handed sextant made by Brandis and Son, New York, circa 1920. The right hand is easily free to record the sight.

1—Hold the sextant with the handle horizontal, arc up and index mirror toward you. While looking into the index mirror, move the index arm until you can see a piece of the arc itself, and its continuation as a reflection on the mirror just beyond the mirror frame.

They should match as though they are continuous with no step up or down. The reflection should be centered about midway vertically in the mirror. Severe mirror misalignment will be immediately apparent; slight misalignment is more difficult to detect. One trick I find helpful is to look first with both eyes, then with one, then the other. This gives one three aspects of this intersection of reflection and reality, and a satisfying feeling of having done everything possible.

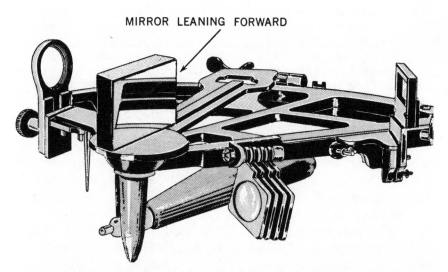

MIRROR LEANING FORWARD

Fig. 2-6 Index mirror perpendicularity check. A step between the actual and reflected edges of the limb indicates the index mirror is tilted. Misalignment will seldom be this dramatic. Courtesy of Defense Mapping Agency.

One consideration: on some index mirrors, the edge of the glass from certain angles of view will reflect a segment of the arc proportional to the thickness of the glass which is not in line with the curve.

Ignore the short segment. We are checking the main reflecting surface of the mirror only. Regard only the big picture. A second caution is offered by navigation instructor Paul Anderton, 'of Annapolis, Maryland, who points out that the validity of this check is influenced by *where* the index arm is set on the arc.

It should be placed in such a position that the mirror reflection is from the extreme opposite end of the arc.

This means the angle of reflection needs to be as broad (close to perpendicular) as possible. When looking into the mirror, if the 120-degree mark can be seen next to the 0-degree mark viewed directly, you know you are spanning the entire arc.

ADJUSTMENT PROCEDURE #1: If the arc looks discontinuous, the index mirror must be angled some to make it parallel with the arc and frame again. Insert an Allen wrench (usually supplied with the sextant) into the hexagonal socket so that it points as near to 12 o'clock on an imaginary clock face as possible. While it is not hard to deduce which way the mirror needs to be tilted to make the arc look smooth, I usually just back off a little (counterclockwise) to determine the effect. On some of the older sextants, a half a turn might be required to cause appreciable motion, but on newer models, a very small turn usually does the job. If backing off was the wrong way, reverse the twist on the wrench and watch the arc and its reflection come into line.

A more accurate way to check the index mirror perpendicularity requires two identical flat sided objects such as machine nuts, dice or dominoes. With the sextant standing horizontally on a flat surface, preferably near comfortable eye level such as the companionway hatch, stand one domino on edge near each end of the scale on the arc. Adjust the index arm until the reflected domino appears adjacent to the directly viewed domino and judge whether the top edges match precisely or not. The beauty of this method is that the line of sight is parallel to the frame and that little segment of edge reflection which was a discontinuous distraction in the other method now falls neatly into line. It also ensures that one end of the arc will be compared with the other, giving a more precise alignment. Since the hands are free to turn the adjusting wrench, the effect of the adjustment can be seen as it is being made.

2—The second check, this for perpendicularity of the frame and the horizon glass, is also quickly done. Set the index pointer near zero degrees and sight on the horizon. If using a scope, readjust its focus for maximum sharpness. With the micrometer drum knob, adjust until the horizon is absolutely continuous. Then lean the instrument from side to side as if it were a pendulum hanging on the line of sight to the horizon. This motion is called rocking or swinging the arc, because if a body is being viewed it will describe an arc in the field of view during the process. *The horizon should maintain its straight line aspect from one side of the swing to the other.* If a step develops in the horizon when the sextant is out of the vertical, or if the horizon splits into two images in the wide view horizon glass, the frame and the horizon glass are not perpendicular. This is called *side error*.

Fig. 2-7 Index mirror perpendicularity check. Top view shows positions of dominos near ends of arc, and index arm in vicinity of 35 degrees. If mirror image of the four-spot domino aligns with direct view of the one-spot domino, the index mirror is perpendicular to the arc and frame (bottom).

Another way to check for the presence of side error when the horizon is dark or obscured is to sight on a bright star with the instrument set to zero.

Turning the tangent screw back and forth past zero should cause the reflected image to move up and down through the direct image. With side error present, it will pass to the side and not right through. *Any side error should be adjusted out as soon as detected.* See **Figure 2-9**.

If the horizon stays continuous during rocking, don't change the setting because you have just aligned for the third check and need only to record the value as read from the scale on the arc. This quantity is *index error*—a measure of the amount the two reflecting surfaces still are out of parallel to each other, provided the perpendicularity of each has been established. It is necessary to check both elements of perpendicularity before checking parallelism to get the index error.

> **ADJUSTMENT PROCEDURE #2:** The screw on the side of the horizon glass frame is turned carefully to eliminate the apparent step in the horizon which is formed when the sextant is leaned sideways.
>
> If the hand on the Allen wrench blocks the view of the horizon it may be necessary to sight and adjust alternately. Make small increment adjustments—no big twists. If the horizon is indistinct, a sharp cloud edge or the sun may be used. As the sextant is rocked, the direct and reflected images will separate sidewise and the adjustment is made to bring them back together in coincidence. A star works well at night (if the sea is not rough) with two images separated laterally to be merged exactly by adjustment.

3—Index error represents whatever total error is residual in the instrument after perpendicularity adjustments have been made. It is spoken of in terms of being "on the arc," or "off the arc," according to whether the index mark falls above or below zero degrees on the scale when the direct and reflected horizons coincide precisely. The mariner's memory jogger when correcting a sight for index error is, "If it's on, take it off. . ." meaning the correction equals the error with sign reversed.

It is neither necessary nor desirable to adjust index error to zero every time one is detected. As long as the error remains at about two minutes or less, it is best to deal with it mathematically.

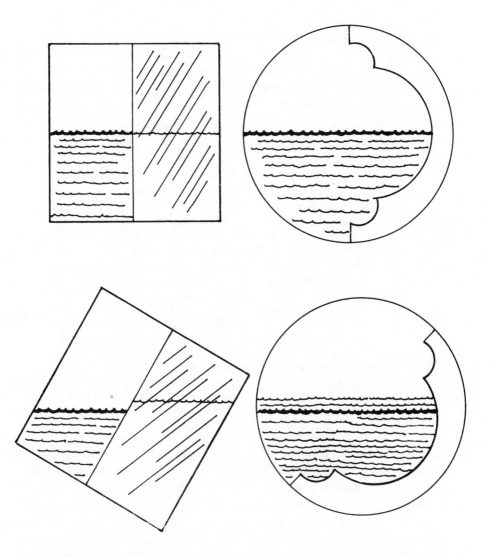

Fig. 2-8 Test for side error. Top: Standard horizon glass, left, and wide view, right, with tangent screw adjusted so that horizon is continuous. Bottom: If a step develops in the horizon when the standard-glass instrument is tilted, side error is present. In the wide view glass, the condition is indicated by a splitting of the horizon line into two images.

Index errors seem to come and go in even the best sextants. One practice that helps reduce the variability of the error is being consistent in direction of adjustment of the reflected horizon when making the match. I prefer to bring the reflected horizon up to the actual horizon because that drives the gear train in the same direction as does bringing down a body for a sight; any lost motion will be the same for both operations. Some people twist the tangent screw back and forth sort of bracketing the horizon and as a result are making a variable out of what could be a consistent, mathematically manageable error.

ADJUSTMENT PROCEDURE #3: The third adjustment, for index error, is the most important. With the telescope sharply focused, make the direct and reflected horizons coincide and note the reading on or off the arc. Then set the scale on perfect zero and the horizons will separate by the amount of index error. Close the gap, this time with the adjustment screw near the top or bottom edge of the horizon mirror until there is one sharp horizon again. Move the index pointer off zero by a few degrees and then recheck for index error. If it is down to within a minute either way, you will probably have a hard job getting any closer. The slightest twist of the screw back will probably overcorrect. The common practice is to make adjustment to index error only when it gets to be over two minutes and then back to less than one minute. Since side error may have been affected by index error correction, it must be checked after index error is adjusted. It sometimes may be necessary to go back and forth a couple of times.

4—The fourth check is for parallelism between the frame of the sextant and the telescope. This is to determine if there is an error of collimation, i.e. alignment. At sea, a gross misalignment can be detected by looking into the index mirror as light would enter from a star. Set the scale on zero. *The reflection in the index mirror of the center line of the horizon mirror should align with the actual center line of the horizon glass and the line of sight should go straight down the center of the telescope.* This is not a highly precise check, but the philosophy is that there is not likely to be a small problem—either a big one from being dropped, for example, or none.

It also is possible to check collimation by sighting directly on a star in the horizon mirror and matching it with the reflection of

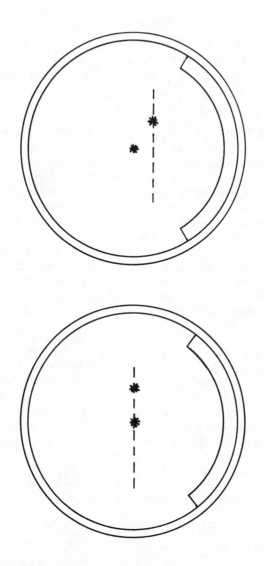

Fig. 2-9 Side error test with star. Turning the tangent screw will cause the reflected image of a star to move vertically as indicated by the dotted line. If the reflected image passes to one side of the direct image, side error exists. In the lower sketch, there is no side error.

a second star brought from more than 90 degrees away by adjusting the index arm. Coincidence should be made near one edge, right or left, of the field of view of the telescope.

If the instrument is then tilted so that the superimposed stars move to the opposite edge of the field of vision without separating, no collimation error exists.

> **ADJUSTMENT PROCEDURE #4:** If the stars separate, there is an error which may be adjusted out by means of a two-screw adjustment at the base of the telescope collar on most sextants. *This adjustment should be undertaken at sea only for compelling reasons and never just to touch up some small and perhaps imagined error. It is really the job of the instrument maker using his collimation range in a shop under controlled conditions.* A level light beam is used to compare measurements at the instrument and at the distant end of a line of sight. Very seldom is there any appreciable misalignment and when collimation error is detected, it usually involves an instrument with an elongated, adjustable position scope such as the Navy Mark II.
>
> Short scopes on mounts not designed for adjustment usually retain their alignment, barring catastrophies such as the sextant being dropped. If the scope is bent and cannot be adjusted, take it off and sight through the center of the collar. This works well and you probably will get sights just about as accurate as with the scope as long as your vision is good enough. When I was a Navy navigator in my mid-twenties and had 20/20 eyesight, I got consistently better star fixes without a scope. The scope serves mainly to provide a sharper image for the navigator; it does not otherwise contribute to the measurement process, but it can detract from it if out of alignment.

There is another way for the navigator to recheck for collimation error which I devised after visiting an instrument shop and seeing how the technician did it. Having determined that the telescope is in true alignment with the frame, either by collimation range or the star sighting procedure, place the instrument on a smooth, flat surface. Sight through the horizon mirror and through the scope backwards at a scale or ruler standing on end on the flat surface and leaning against the eyepiece of the scope. The best kind of scale is a transparent plastic scale that is marked in millimeters. With a light source and a white surface behind, it will be in focus and legible from close up, if looking through the 4x40 star scope. Looking backwards through a 6x30 prismatic scope, the scale will come into focus if viewed from a greater distance. In either case, the object is to establish a benchmark figure for the distance from the flat surface to the top of the pupil lens field of view.

Note it down in some enduring spot, such as the margin of the instrument's certificate in the lid of the box. If the figure changes, the scope has moved and no longer is parallel to the frame.

*   *   *

Adjusting the index mirror and horizon glass to remove errors detected is a task some navigators approach with sweaty palms. It needn't be feared if one rule is kept in mind: If there are two opposing adjustment screws on a glass, one must be loosened by backing off before the other is tightened. The glass will not bend and, being brittle, will snap quite readily if not provided room in which to move.

Having given that grim warning, I should at once add that the good news is that most modern sextants use a better system for holding the mirrors eliminating the two-screw, glass-cracker system. This uses a single screw acting against spring tension, which makes adjustment easy and sure. The horizon glass may still have two adjustment screws but they are not opposing—one bears on the top or bottom edge and the other on the outer edge. Each screw tilts the glass in a different plane. I suppose it is still possible to snap a glass with these, but it is not likely since the spring's tension yields as the adjusting screw presses harder against the glass.

CHAPTER 3

ATTACHMENTS
AND ACCESSORIES

Buying a first-rate sextant is getting more like buying an automobile in that there are many options to consider. Some seem highly desirable, while others are of questionable value.

The following is a list of major attachments that may be considered.

ASTIGMATIZER

An astigmatizer is a lens that is attached to the sextant as part of the group of sun shades that pivot in or out of the line of sight or light ray path in front of the index mirror.

It replaces one of the shades on some installations and looks very much like a shade itself. In operation, it converts a speck of starlight into an elongated horizontal line of light across the field of vision. The idea is that a flat streak will be more easily matched with the horizon than a small point of light. The light remains horizontal and moves diagonally across the field of vision, scuttling crab-like toward or away from the horizon. Eric Hiscock, in his book *Come Aboard*, praises an early type of astigmatizer, called a stellar lenticular, for its ability to show when the sextant is vertical by providing a reference line to keep parallel to the horizon while sighting.

The astigmatizer I used on a Cassens & Plath sextant produced an elongated streak of light. The streak, however, did not tilt degree for degree as the sextant was tilted, and was of little use as a vertical reference. The Big Dipper looks like an upset bar graph when viewed through an astigmatizer, while the sun appears vaguely like a squashed marshmallow. Astigmatizers are not cheap, costing $175–$200.

I would recommend trying out a borrowed one before investing.

WOLLASTON DOUBLE STAR PRISM

This is another device that goes on the shade rack between the index mirror and horizon glass, or on the front end of the scope, and is used for altering the image of stars to render them easier to measure.

It refracts a star into two points one over the other.

When the horizon is faint or ill defined, it is supposed to be easier to place the two images fairly astride the horizon, one over and one under by equal distances. I have had the opportunity to use one only briefly and then the horizon was disappointingly sharp. I did think, however, that it might prove useful in bad conditions, if they are still in production; they may be extinct.

WIDE VIEW HORIZON GLASS

Introduced in the U.S. in 1980, the wide view horizon glass is relatively cheap, if bought with the instrument, but relatively expensive if put on later. This option, also called a beam converger by Davis Instruments, is distinctive because it has no mirrored opaque right half. You can see through the entire horizon glass. At the same time, because of a coating of quartz crystals, the entire surface is reflective. Davis Instruments holds the 1979 U.S. patent for the process and at one time made horizon glasses for other sextant companies. An early-1900 C. Plath catalog shows a device that operated on the same principle.

There is some controversy over this attachment concerning whether the horizon is obscured to an unacceptable degree by the coating. It generally is agreed that for sun sights, where the horizon is usually strong and sharp, there is no problem. At twilight, when the horizon is faint, any obscuration at all may

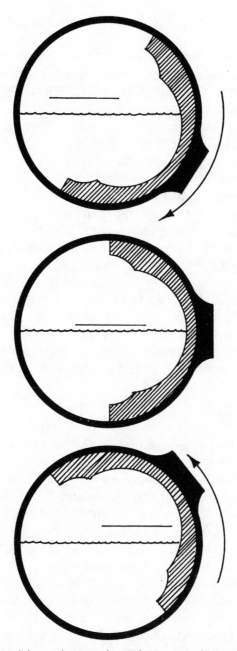

Fig. 3-1 Viewed through an astigmatizer, a star is stretched out in a horizontal line that stays nearly level and simultaneously moves diagonally across the field of view as the sextant is rocked (top and bottom views). In the center view, the sextant is being held upright. The star must be merged with the horizon for the sight; however, the flattened image often tends to be obscured.

increase the difficulty of sighting. One school of thought is that people who don't ever take star sights in twilight conditions need not concern themselves about this and should get the wide view glass. Avid star shooters may want to think carefully before adding this option. About a third of all new sextants are sold with the wide view glass; however, it also should be noted that of those, about a third are returned for conversion to standard horizon glasses. The cost of changing your mind is about $110, while adding the option at a later date can be more than $250.

My experience with the wide view glass includes periods of weeks of star sights and I don't think I ever could see the horizon noticeably better bare-eyed than through the sextant optics and horizon glass. It may have been that the dimmer horizons would have looked sharper through the clear side of a standard horizon glass, but I never felt handicapped.

The single biggest disadvantage of the broad view horizon glass is the loss of the vertical reference that the edge of the mirrored half provides on the standard glass. Looking through that wide, round window of the wide view, one cannot tell if the sextant is being held upright or slanted. Swinging an arc to test for verticality becomes vital, and, if the vessel is making considerable motion, difficult as well.

PRISM LEVEL

This attachment is clamped on the rectangular horizon mirror of the Davis sextants and also comes in models made to fit other makes such as Weems and Plath, Cassens & Plath, C.Plath, Tamaya, Freiberger, Simex and Navy Mk III. It is a C-shaped clamp that grabs the left side of the round horizon glass and holds a precision coated glass prism in the edge of the line of sight to the horizon—if using a 4X40 or less powerful scope.

With the Davis Prism in place and aligned, the horizon will be seen as a continuous line only when the sextant is exactly vertical. Tilt the instrument and a step or notch forms and two horizons develop. The need to swing an arc is eliminated. A celestial body can be centered simply by moving left or right and the verticality of the sextant maintained. Some say they find it difficult to look at both the body in the center of the horizon glass, and the left side of the glass to see if a step has developed. I find

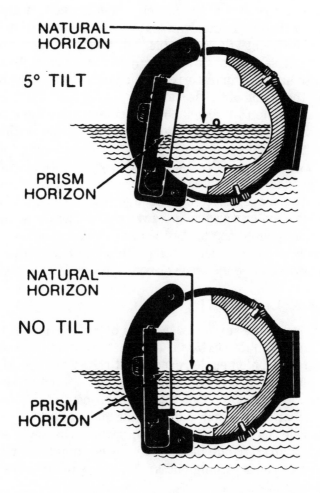

Fig. 3-2 Davis Prism Level. Attached to the left side of the horizon glass, the Davis Prism Level gives a positive indication of whether the sextant is level—a critical requirement for accurate sights. Drawing courtesy of Davis Instrument Co.

this to be no trouble at all. While concentrating on the body my peripheral vision tells me if the horizon becomes discontinuous.

The Davis Prism is easily attached and aligned. A cross hair target is attached temporarily in place of the scope and the instrument stood on its legs flat on a table. The target is sighted through the horizon glass and the prism is adjusted with a knob until the target cross hairs are undistorted, which ensures that the prism is parallel to the frame of the sextant.

It is advisable to recheck this alignment from time to time. Temperature changes could alter the alignment since the frame of the prism is made of plastic.

The Davis Prism to fit sextants by Tamaya, C. Plath and Cassens & Plath sells for about $43 and impresses me as a useful accessory which should make a marked difference in sight accuracy, particularly for those who find swinging an arc awkward business—and many do. The level can be used on standard horizon glass sextants as well, although the need there is not as acute. They don't work with a 6X30 scope, though, because the prism doesn't show up in its field of vision.

BUBBLE ATTACHMENT

The first bubble sextant was made by Mr. Hadley in 1733. In recent times, bubble sextants are most commonly used in air navigation and some people reason that the same type of instrument ought to be good for yacht navigation too since both platforms move around a lot. This theory is wrong on two counts; the kind of airplanes that use bubble sextants provide a whole lot smoother ride than any yacht or ship. Second, the lower accuracy of which the aircraft bubble instrument is capable may be good enough for a machine going 500 knots, but is inadequate for one making 5 knots.

The alluring feature of the bubble sextant is that it does away with the need for the horizon as a reference point. Instead it uses the principle of a bubble level to establish the horizontal plane. Because the horizon sometimes is vague and dim and often altogether invisible, as at night, it would be a fine thing to be independent of it. Then star sights could be taken all night long, avoiding the time pressure generated by the brevity of twilight. Many navigators would like to try out a marine bubble attachment but are deterred by the cost—$990 or thereabouts for a Plath—more than many sextants.

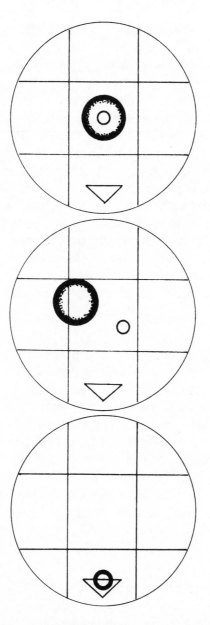

Fig. 3-3 Bubble sextant field of view. In top drawing, the sextant is aligned on the sun and is level—the situation necessary for taking a sight. The sun is centered in the bubble, which is centered in the grid. In the center drawing, the sextant is cocked, that is neither level nor plumb. It is not enough to make the sun and bubble meet fleetingly in the center of the grid. The bubble must be at rest there. The bottom view shows the sextant elevated to put the bubble on the triangle, its required position for adjustment in size.

Why they are so expensive is a mystery. They consist of a low powered scope arranged to look through a container of syrupy liquid such as glycerine in which there is a bubble. The bubble size can be controlled by adjusting a knob which squeezes a bellows to increase pressure on the bubble. The bubble is illuminated by a light bulb powered from the sextant's handle battery compartment—without the light the bubble cannot be seen at all. When the bubble is precisely in the center of the field of vision, the sextant is precisely vertical. A navigational body which is being held centered in it when the bubble is centered in the field will be measured for altitude without reference to the horizon. The trouble is that every wobble of the vessel causes the bubble to scoot off to the left or right and, as if that were not bad enough, also moves the line of sight off the star.

NECK STRAPS

While a lanyard may save a sextant from a hard knock if dropped, a neck strap will do that and more. It will provide a safe and convenient place to stash the instrument between sights. A camera neck strap can be adapted for the job. The trick is to attach it to the instrument at two points—the aft end of the arc and the leg at the top of the handle—so that the sextant rides in a controlled manner on the navigator's stomach. I attach the metal rings that come with the camera strap to the sextant frame with nylon wire ties. It is great to have two free hands and a safe, convenient instrument while awaiting Local Apparent Noon (LAN), or recording a sight.

VISUAL AIDS

Probably the toughest situation the navigator faces, second to total cloud cover, is the horizon being ill defined or deceptive. The best horizons are obviously had during daylight hours and there are many days when the moon as well as Venus can be sighted with relative ease. The horizon often becomes elusive during twilight hours—the best time for star shots—which is why the navigator who ignores the opportunity for a sight of a daytime moon or Venus is imprudent indeed.

The horizon is still there at night and the ability to use it all night long would be a marvelous advantage, but this is not al-

Fig. 3-4 This neck strap is a camera strap and bears the weight of the instrument comfortably. It differs from the usual emergency lanyard in that it is attached so that the sextant rides serenely against the stomach and can be kept there indefinitely, which frees the hands between sights.

ways possible. A moon illuminated horizon can be used, but the navigator must be careful not to be tricked. More on this in Chapter 6.

It is also possible to use a moonless night horizon. An omen that a good night horizon may be coming is an unusually sharp daytime horizon associated with high barometric pressure and cooler temperatures. If the daytime horizon has been indistinct from haze in warm temperature conditions, the night horizon is apt to be tough to find as long as the same conditions prevail.

Given favorable conditions, the navigator must deliberately adapt his eyes to the dark to get night sights. World War II submarine navigators wore dark red lense goggles for an hour or so before the vessel surfaced, allowing them to pop out for a quick round of sights. Good fixes were found to be possible with the very faint star-illuminated horizons which could only be seen with well adapted eyes.

I have used red goggles with good results in dark adaptation and the goggles need be nothing fancy. A pair of ski or motorcycle goggles with red celluloid from the stationary store cut to fit the frames will work fine. Add multiple layers for denser protection and experiment to see how much shading your eyes require. The submarine navigators may have been enduring dark adaptation for longer than necessary. Dr. Robert Fornili, of the U.S. Naval Academy Eye Clinic, says total adaptation may take up to an hour, but that as little as 20 to 30 minutes will bring a person's eyes very close to that point. The sensitivity to faint light is increased as much as 100,000 times. Think of what that short period behind red lenses can do for your ability to distinguish a darkened horizon. The tricky part is to completely avoid seeing any nearby light and ruining the adaptation.

The other night vision improvement might be described as a half measure. G.D. Dunlap, in his book, *Celestial Navigation with HO 229*, (International Marine Publishing Co., Camden, Maine—1977) suggests wearing an eyepatch over the sighting eye in order to dark adapt that eye separately. Eye patches are available from some drugstores and hospital supply places.

The adjustment of the retina of each eye can proceed independently, but when using the dark adapted eye, the other should be closed, which is a natural thing to do when looking through the scope anyway.

In taking night sights, it is helpful not to strain too hard and to maintain a deliberately relaxed attitude. With time pressure removed, there is no hurry and it is best to take several sights of each body and average the time and altitude, first discarding any obvious misfits.

The night horizon being a mere differentiation between two grades of blackness will never be as clearly defined as the daytime horizon to which color differences between sea and sky contribute. The thing to avoid is staring and straining to find and adjust the juncture of celestial body and dark horizon with exquisite precision as if working with a lighted horizon.

The trick is to say—"There it is; bring body down; Mark." Record the data and go on. Intense exertion over night sights will only cause a headache.

CHAPTER 4

CARE, MAINTENANCE
AND REPAIR

T he accuracy of the sextant depends directly on the precise relationship between one revolution of the tangent worm screw and an advance of one degree of the index arm along the arc.

Excessive wear of the screw or the teeth of the limb will alter this relationship. So will accumulation of dirt and grit in the grooves that are the teeth of the gear rack. That's why regular cleaning and lubrication of the sextant is important.

But don't get carried away. It is ironic that those who are most careful of their sextants—and this includes instrument repair and adjustment people who should know better—sometimes unknowingly mistreat their instruments by shining the face of the arc until it sparkles. Brass polish is grit and grit grinds gears and other metal-to-metal parts. As far back as 1881, Squire Leckey, in *Wrinkles in Practical Navigation*, warned that one must give positive instructions to the repairman to not polish the arc, ". . .otherwise he is certain to do it unbidden."

Fortunately, modern sextants can take this mistreatment better than the older vernier models. They have deeply machined graduations on the arc that are relatively unaffected by abrading brass polish. Older sextants, however, have finely engraved inset silver arcs that will not take much shining before becoming unreadable.

The right way to clean the arc is with ammonia and a cotton-tipped swab. This will remove the grime and lighten up the brass face of the arc without doing any damage. Several applications may be required and some rubbing with the swab. Be careful to keep the ammonia off the rest of the instrument.

The grime in the teeth on the limb is loosened by ammonia and can be brushed out with the small stiff brush that comes with the sextant or, if that is missing, with an old tooth brush. The limb may not look as dazzling as after brass polishing, but it will be in much better condition—clean and smooth, with no gritty crevices, and easy to read. This process, of course, will de-lubricate the sextant completely, leaving metal to metal contact of the sort that would ruin an engine in seconds.

What to use? Not 3-In-1 Oil!

In the old days, before The Endangered Species Act and the Marine Mammal Protection Act, sextants were customarily both lubricated and protected by head melon oils of the smaller toothed whales. This was a high grade whale oil that had characteristics making it superior for instrument use. The oil had a high viscosity and equally high proportions of the right sort of molecules to give it superb resistance to both oxidation and gumming. Its triglycerides produced superior film strength on metal surfaces and it was ideal for both protecting and lubricating the instrument. This oil is no longer available in the U.S., but there is a superior substitute.

William F. Nye, Inc., of New Bedford, Mass., one of the leading producers of instrument oils in this country, says the state of Arizona filled the oil gap with a "weird, almost spooky fluke of nature," when farmers there began growing *Simmondsia Chinensia*, otherwise known as Jojoba oil. Squeezed from little bright red beans, Jojoba oil is molecularly similar to fine whale oil and has one distinct advantage—no smell. Nye added synthetics to Jojoba oil to overcome its deficiencies and the result was Clock Oil 140-B.

The oil is thinner than one would imagine, but the tenacious surface film it forms acts like wax on a car in making water stand up in little separated globules. Nye sells a one ounce bottle for $3.50 (see appendix for address), while pure Jojoba oil can be bought at health food stores for about $4.50 an ounce, which is a considerable savings over mail-order catalog prices.

Start lubrication by turning the sextant upside down and oiling the tangent screw with it set at the low end of the arc, then screw it along the entire length of the arc so that it will lubricate each tooth of the limb in succession. Put a hint of a drop, like what a dipped toothpick (thin end) will hold, on the head of each adjustment screw and wipe away excess.

These screws tend to corrode and stick so that the instrument cannot be adjusted properly. Put a drop or two as near to the pivot pin in the top of the index arm as you can get. This often means in a crack above the pin where it will run down to the pin itself. Wipe away the excess with a small absorbent cloth, keeping clear of the mirrors and handle. Wiping the entire frame with this oily cloth is probably beneficial. The frame also can be coated with a silicon moisture displacer like WD-40, which has no appreciable sticky component. Be careful to not get any of these oils or sprays on the optics.

One must use a little care in applying the aerosol WD-40 or any of the other moisture displacers that come in spray cans. The spray really messes up the mirrors and optics and is difficult to clean off. Before spraying, it is best to encapsulate the mirrors and shades with sandwich bags properly stopped off with twisties at appropriate choke points. This is not too troublesome and much easier than cleaning the glass surfaces later.

Dismount the telescope beforehand.

If a sextant has been really wet down with saltwater, the best cure is holding it under a running fresh water faucet. Lacking this convenience, the next best choice is to immerse the instrument in a bucket of fresh water.

To repeat, *don't forget to remove the scope!* Also, don't forget to remove batteries from the handle and re-tighten the compartment cap securely. Dunking sounds like drastic action and it is, but a coating of salt is bad enough medicine to warrant such measures.

Give the mirrors and wiring special attention when drying. Remove light bulbs and dry the inside of the sockets carefully. After drying the entire instrument with a cloth as thoroughly as possible, arrange for a breeze and a low humidity environment, if at all possible. A natural breeze and sunlight, a fan in a heated space, an oven preheated to a low temperature (and then turned off), close proximity to an air conditioner duct—whatever can be contrived will serve. If the sextant only got a little spray, a thorough wipe down with a damp cloth frequently rinsed and wrung is sufficient.

In all cases, re-lubricate.

While you've got the WD-40 at hand, unscrew the battery compartment cap on the end of the handle and remove the batteries. Spray the inside of the battery compartment. This is beneficial in a couple of ways. It protects interior surfaces against corrosion, improves conductivity of electrical contacts and even makes the screw cap work better. Just don't make it dripping wet inside; that won't hurt anything except that batteries will get wet and you'll get the lubricant on your hands when you remove them.

On the subject of removing batteries, I have formulated the following general philosophy: If you are forced to use the cheap kind of graphite batteries, which last but a short time and leak acid thereafter, you had better heed the manufacturer's advice and remove them every day. If they do leak and swell unattended in the sextant handle, you may have to get a new handle. The modern solution is to use rechargeable Nicad batteries which don't leak. Since they must be removed to be recharged, they get inspected periodically. Have two sets, one in the sextant and one in the charger.

Remove even Nicad batteries during extended periods of non use since they will dwindle to lifelessness anyway.

Most sextants have three legs on which to stand when not being held. Two of the legs are near the ends of the limb on the handle side of the frame and the third leg is near the top of the frame and usually joins the top end of the handle. The three legs are of equal effective length so that the sextant frame, in repose, is level with the tender mirrors on top. On a smooth surface such as a chart table, these metal legs start to skid at an inclination of only about 10 degrees.

I have determined this with my protractor and a tilting board.

Ten degrees is not an unusual amount of roll on shipboard. The antidote? Hardware stores sell neoprene washers for use in faucets. Flat on one side and cone shaped on the other, they are ideal for sticking on sextant leg ends for anti-skid purposes. My test board can be tilted to almost 40 degrees before the slide begins with the washers in place. Put the washers on flat side down.

If the fit is not snug enough, a judicious drop of silicon rubber cement from the bottom side will keep them in place.

This brings up the subject of where to put the sextant when it is not actually in use.

The old rule that it is either in hand or in the case at all times may be best for the instrument, but often is inconvenient. During unscheduled interruptions or emergencies, few of us have the poise to re-stow the sextant before panicking. Some people just put it down on a flat surface such as the chart table or cockpit seat. Merchant ships often have little shelf writing tables behind the windscreen. Only about 15 inches square, this convenient table is an almost irresistible lure for stashing sextants. I have seen unattended sextants perched there many times, awaiting only a 10 degree roll before plunging four feet to a steel deck and disability retirement. Far better during urgent interruptions to put it on the deck or cockpit sole in the first place—as long as it is out of the way and won't get stepped on.

A better preplanned solution for temporary parking between sights as well as emergencies is the sextant hook. This often is made from a welding rod bent in an S shape. One of those galvanized S hooks purchased from a hardware store will do as long as it is a big one. Coat hanger wire is inadequate and cannot be relied on to support the weight of a standard instrument—and it easily rusts. On most ships there are armored hook and support brackets all over the overheads of the bridge and chart house.

I like to put up a hook in a corner near each door to the wings of the bridge and a third over the chart table. The sextant is hung by its heels—that is with the hook through the frame at the zero end of the limb. It can be hung in any attitude so long as the wire runs clear of tender parts. A piece of rubber tubing can be slipped over the lower end of the hook so that the finish of the sextant will not get scratched. Be sure the sextant cannot swing into anything as the vessel rolls. This is a very secure place for a sextant. Of course, the location of the hook should be chosen to minimize collision potential from passers-by.

Fig. 4-1 Sextant hook. Here the instrument swings from the outhaul cleat of the main boom out of harm's way and yet accessible.

On a sailboat where many things are more difficult, it may be hard to find a good place for the hook to hang. From the overhead in the aft end of the cabin on the quiet side of the companionway where it can be reached from the cockpit, but is out of the weather, is good. A hook can be put over one of the pipe frames of a cockpit dodger to keep the instrument in a protected position. I can almost guarantee a dry mouth and quickened pulse when first suspending the treasured instrument thus. Although it appears precarious, it is really more secure than when in hand, being physically attached and free to hang clear. Only make sure the depth of the hook end is generous—three or four inches—to guard against hitting a pothole in the road; you want to make certain the sextant can swing without banging into anything.

Of course, you wouldn't want to hang a sextant in the rigging if there were any spray or leave it there for an extended period in direct sunlight, but it is an ideal arrangement while awaiting twilight or the sun's crossing the meridian.

There is no disputing that the very safest place for the instrument is in its nest, but some of them have faults too. One is the situation in which the two legs on the limb do not touch the bottom of the box when the handle is latched into its recess, allowing the sextant to rock back and forth slightly. Sponge rubber pads glued to the bottom of the box where the legs strike cure this condition. The other common box shortcoming is not so easily cured. The index mirror shades pivot into and out of the line of sight depending on the amount of shade needed for a given body. This can vary from none for a star to plenty in the case of the sun in the tropics. Folded part way out, those shades can sometimes hit the side of the box. This is bad for the shades and the box gets gouged by the handles on the shades. The only solution I have found to this little problem is cultivation of the habit of folding the shades inward before the instrument is put into the box.

Talking about sextants fitting the boxes, it is a great blessing if the box will accommodate the instrument with its scope in place for sighting so that the sextant is ready to use from the case. It is a pain to have to attach the scope after the sextant is removed from the case because it won't fit in with the scope on. If I were choosing from otherwise equal instruments, that feature might make up my mind. Another great convenience is accommodation for a spare scope—a recess or clip. It is not good to have the second scope loose inside the case to rattle around and perhaps damage a mirror. This convenience is easy enough to make yourself if the box is not already equipped.

What is to be the disposition of the box with the sextant secure inside? In a yacht it is often wedged in a locker under a bunk or perhaps on a shelf above the cushion backs of the saloon seats, or in some other more or less secure niche. The trouble with many of these improvised solutions is that it's so much trouble to get the case in and out of its secure spot. The best solution aboard a yacht is to attach the sextant box itself to a bulkhead or other convenient vertical surface. There is no need to build a rack to hold it. In fact, there is an advantage in not having a rack—the instrument is more accessible. The box is screwed directly to the bulkhead, if it is substantial enough for screws, or bolted through if it is a thin partition. The case should be positioned handle up with the bottom of the case against the bulkhead and the top opening outward and downward. The top open support bracket with which most cases are equipped will support the opened lid somewhere near horizontal where it is useful as a temporary shelf for gear connected with taking sights such as a pad and pencil.

A person who can't stand the thought of drilling holes in the bottom of the sextant case can avoid that with a little ingenuity. A portion of any suitable flat area can be fenced off with wooden pieces about half as high as the depth of the sextant box so that it will just fit in and be held securely. Be sure there is enough space between the fence and the bulkhead behind to allow room for the top to open.

Sextants are really very stable, durable, low-wear factor devices that have an inherent tendency to continue to operate satisfactorily right up to the moment of a catastrophic incident and often even beyond. I have used more than one ship's sextant with a crooked leg or battered limb end—evidence of having been dropped, which still worked fine. Sextants are not subject to mysterious ailments like electronic equipment. If the instrument is properly adjusted, it will function correctly unless badly mistreated.

I may as well admit here that I've never dropped a metal sextant and can't tell you from experience how they bounce, and

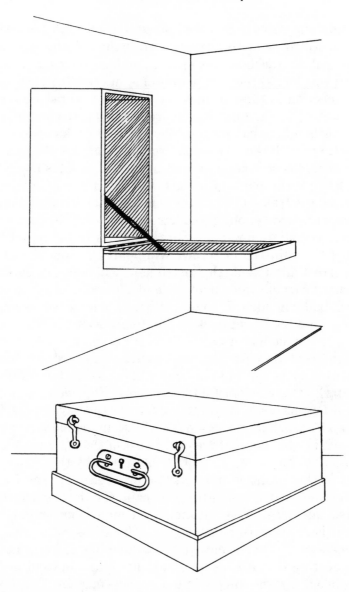

Fig. 4-2 The sextant box can be screwed to a bulkhead or other vertical surface in an out-of-the-way location. When closed and locked, it cannot be removed without violence. When open, the top makes a convenient temporary shelf. On a flat surface, a fitted fence will secure the box for anything short of pitch poling, and a bungee cord over the top will help in the event of even that.

as Mr. Budlong notes in *Sky And Sextant*, they don't float well at all. If you still are able to retrieve it after the fall, the most likely damage will be cracked index mirror and horizon glass. Protruding from the frame, they are vulnerable in their brittleness and firm connection to their frames through their adjusting screws. Fortunately, they are easily repaired if using the modular exchange technique; that means take out the broken pieces and replace them with the spare you brought along for the occasion. Adjust out side error and index error in the usual fashion (Chapter 2). Even if the point of impact was the frame of the broken mirror and the frame is bent, there is a chance that the range of adjustment is wide enough to allow you to bring the instrument back into visual alignment. Even though its frame may be distorted, if the adjustment screws will hold the replacement mirror in correct alignment, the instrument will measure correctly. If the mirror frame were bent beyond the scope of adjustment, and a fix badly needed, I would not hesitate long before putting the pliers to work to get back within tolerances.

How many navigators actually carry a spare set of mirrors, one might ask? Very few, I would guess. Some sextants used to come with a spare set, but only one still does that I know of—the Navy Mark III by Scientific Instruments. Replacement mirrors for the other makes of sextants are fairly expensive items. They are about $70 each for a Cassens & Plath instrument, for example, which is a big percentage of the cost of the instrument. It is interesting to note they are not stocked even by the big sextant dealers in this country and must be ordered from the factory. This signifies that there is little demand for them because of a low breakage rate. If you examine an index mirror out of its frame, it is impressive because of its thickness, which is about $3/16$ inches—somewhere near that of automobile safety glass. I'm not suggesting that it is as tough as auto glass, but neither is it as fragile as, say, the mirror in a woman's compact. The sextant mirror has to be fairly tough because of the way it must be mounted on points so that its angle with the frame can be adjusted.

The common problem with mirrors is deterioration of the silvering so that reflections of bodies gradually become indistinct and finally in advanced cases imperceivable as the silvering flakes off. Salt water atmosphere speeds the process. Salty moisture entering behind the mirror causes dark spots to appear, which gradually widen and spoil the reflectivity, even before flaking becomes evident. This process can be greatly retarded by proper stowage, but sooner or later, ordinary use will cause the silvering to go.

If the decision is made to carry a spare set of mirrors along, they should be wrapped up and sealed in an airtight container so that their silvering will be preserved.

Fortunately, resilvering is inexpensive and takes but a few days. Maryland Precision Instruments, in Baltimore, charges about $60. If your need is urgent, they'll give you already-silvered mirrors in exchange for your mirrors and the cost of the job—assuming your mirrors are in good shape. The service is not that good everywhere, so the idea is to get the job done in a timely fashion—that is, long before you are forced to the conclusion that it is necessary. Perhaps a guideline for when to have it done will help.

If, with the aid of a magnifying glass of suitable power or through the instrument's own scope dismounted and reversed and under a strong light, any spot whatsoever can be detected in the silvered surface on the back of the mirror, the time to resilver has come. Don't try to squeeze a little extra mileage out of the mirror unless you are mid-ocean at the time. It doesn't matter if the spot is tiny and way over to the side and out of the way. My current number one sextant, with moderate use, took four years to develop that first tiny spot. Though the rest of the mirror still looked great, I had the mirror resilvered.

It is possible to do your own resilvering; however, it is both difficult and dangerous because of the chemicals involved. Better to have a second sextant aboard. Should a real emergency arise, do as the mariners of the middle eighteenth century did—paint the back of the mirrors black.

Previous to then, sextant mirrors had been made of highly polished metal, usually steel, and were called *specula* (plural), or *speculum* (singular).

Silvered glass mirrors were around, but they had a tendency to introduce distortion because of the difficulty of making both surfaces absolutely parallel. Then in 1765, the Rev. Dr. Nevil Maskelyne, father of *The Nautical Almanac*, said that to avoid the problem, just grind the back surface of a piece of glass flat and paint it black. The black absorbs light that reaches it and the reflection seen is from the front surface only. It works. Stars can be seen clearly enough and the sun perhaps even better than from a silvered surface. I blackened the mirrors of one of my sextants and had no particular difficulty getting good sights.

The biggest problem with doing this is getting the old silvered surface with its attending paint or varnish seal cleaned off completely. One of the best ways to do this is with simple paint remover such as Savogran Strypeeze Paint and Varnish Remover which contains methanol, toluol, acetone and methylene chloride. It is dangerous stuff indeed.

The removal process may take days. If necessary, scrape to help things along, but use nothing harder than a piece of wood to avoid scratching the surface of the precision-ground glass. The best and cheapest mirror silvering solvent I found is muriatic acid, which is used for cleaning brick work and porcelain. It is available at hardware stores for $1.50 a quart. It must be handled carefully, in accordance with the very explicit advice on the label.

An ounce or so in a jar will bathe the silvering off clean over night without any rubbing at all. After the silvering is off, the glass should be cleaned with detergent and water till squeaky clean.

There is no trick at all to applying the paint. Paul Anderton, my Annapolis navigator/lawyer friend, suggests a special paint called Pactra, which flows onto glass with extraordinary smoothness and gives a highly reflective surface. You can buy this at hobby shops. Rustoleum flat black works well as do high gloss automotive enamels. Probably any black paint will work well.

The application is simple. Take a soda straw and insert it an inch or so into the paint. Put your finger over the top of the straw and lift the tiny paint load to the mirror. The small blob of paint, when released by lifting your finger, should flow slowly over the entire surface and not leave too much to bead up around the edges. If the straw is not available, use a cotton swab for a brush.

If doing the horizon glass, mask off the side to be left clear with tape. After the paint is dry, don't pull the tape off impulsively lest a ragged edge be left. To ensure a sharp edge, use a single-edge razor blade guided by a straight edge to score a line in the paint adjacent the edge of the tape. Then carefully pull the

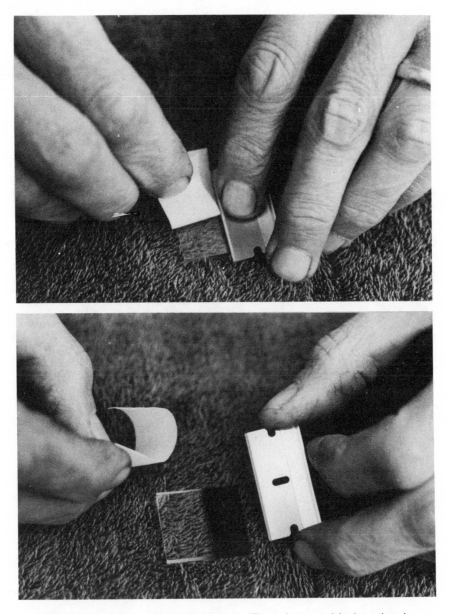

Fig. 4-3 Horizon glass unmasking. Top: A razor blade edge is used to hold down the new black paint back surface while the masking tape protecting the clear side is peeled off. Bottom: Photo shows the finished replacement horizon glass with sharp edge between clear and reflecting portions. Photos by Nancy Bauer.

tape off. The quality of the edge is not that important, but a sharp line is good for the morale. Give the black paint a protective coat of cellophane tape.

Testing the accuracy of the black mirrors, after alignment as described in Chapter 2, is easily done by measuring the horizontal diameter of the sun. Measuring the sun with the sextant on its side eliminates refraction errors that creep into a vertical measurement. Starting with the sextant set at zero, separate two images of the sun and touch their limbs together delicately. Add or subtract IC as appropriate and the result should be exactly twice the semi-diameter (S.D.) listed at the bottom of the column for the sun in *The Nautical Almanac* page for the date. If it is not, the difference should be combined algebraically with IC and subsequently factored into each altitude taken.

Think of them as the Black Mirror (BM) correction and the combined form as an ICBM.

CHAPTER 5

HOW TO BUY A SEXTANT

WHETHER AND HOW
TO BUY A USED SEXTANT

In January of 1983, I ran a classified ad in *Yankee Magazine* announcing I would pay cash for a used sextant.

The ad cost me about $75 to run once. *Yankee* was selected because it is not a boating magazine, has a large circulation (about 850,000), an older audience and the typical subscriber is allegedly frugal. I was aiming for people who might have a sextant they didn't want. Because of the seafaring traditions of the area, I reasoned that there might be a vast number of sextants in New England attics.

About 20 responses arrived over the next 12 months. Some sent pictures and full descriptions; others said something like, "I got this old thing I think is what you're looking for." If an instrument sounded good and was a known make, I offered $100 on condition I could return the instrument if it turned out to be unusable. I sent instructions on packing, shipment, insurance, and agreed to pay COD shipping costs.

I have bought several sextants as a result of that one ad.

One is an old rosewood octant that may be of historical value.

All those I bought were worth more than the $100. I sent only one back—it was a bubble sextant of the aircraft type. At the time, I had not yet read John S. Letcher's *Self Contained*

Celestial Navigation with HO 208. If I had, I might have bought the sextant just to test his assertion that it can be modified and used as a non-bubble instrument with good results.

I offered $25 for a plastic lifeboat sextant and got just about what I paid for. I learned later that Davis sells them new for the same price (but without case). I was offered four instruments owners felt were worth more than my $100 limit. They seemed knowledgeable about what they had. The instrument that sounded best was a C.Plath about 20 years old whose owner seemed offended by my offer and said he wanted $300.

Chances are it was a good instrument worth that much.

I don't subscribe to the old maxim of maritime philosophers about never buying a used sextant. The implication is that using them ruins them. This is untrue. Sextants wear out very slowly and they usually get treated with utmost gentleness and even reverence. Damage is not hard to detect if the instrument is examined intelligently. I think I would modify that old rule to read: Don't buy a used sextant for more than half the retail cost of the least expensive, new, full-sized instrument—unless the money saved is critical to your budget. This would figure out at $450 as the most that would be paid for a used instrument unless there are irresistible inducements such as being in really fine condition, coming with extra accessories, having antique value and so forth. Don't let anybody convince you that his 20-year-old German sextant with small mirrors is worth $500 or $600. Some old, retired ship masters tend to have particularly inflated ideas of the value of their faithful old instruments.

There are many good used instruments around and by exercising caution and avoiding hasty deals, one can end up with a sound instrument for a lot less money. Here are some ground rules and points to check to avoid getting saddled with an inferior instrument.

1. Do not consider any instrument without a micrometer drum and tangent screw. The inconvenience and error making potential for the old slide vernier arrangement are intolerable for the modern navigator.

2. Evaluate the size of the mirrors. One of the best improve-

ments in modern sextants has been the enlargement of the index mirror and horizon glass.

The index mirror should be 1½ × 2 inches, and the horizon glass two inches in diameter. This makes for fast acquisition and better control of the instrument. Visibility through the instrument is better. Don't buy an instrument with smaller mirrors. Incline toward larger mirrors in deciding between two otherwise equal instruments. Small mirrors can be replaced by larger ones on some sextants, but the costs are substantial. Incidentally, bad silvering on mirrors is a minor problem although it makes a useful talking point when bargaining. A mirror resilvering job is inexpensive.

3. A serious condition that may be hard to detect is what Captain Oswald M. Watts, in his book *The Sextant Simplified*, calls "The tormented sextant." He describes this as an instrument that has been adjusted so frequently that the adjustment screws no longer work properly. If the screw holes are enlarged or puffy with corrosion, it may be difficult or impossible to keep the instrument in adjustment. A buyer should examine each adjusting screw head and hole with a magnifying glass. Look for enlarged holes, spoiled slots and rounded nut facets. If any look questionable, remember to point them out later if you decide to take it to an instrument man for his opinion.

Fortunately, tormented sextants are not common. Usually people are reluctant to adjust them. A small index error—up to about two minutes—should be dealt with mathematically and not continually be adjusted out. There really is no good reason to overadjust a sextant.

4. Unscrew the cap to the battery compartment, usually in the handle, and examine the inside.

If the sextant has been put away for a long time without removing the batteries, there is a fair chance they have leaked. Since the compartment is supposed to be watertight, that could happen with no outward sign and often will make a replacement handle necessary. This is not too expensive, but should be a negotiating point. If the compartment turns out to be clean, put in batteries and test the circuits.

5. Tell the owner that you wish to try the instrument and perhaps have it examined by an instrument person. This is one

advantage enjoyed by the used instrument buyer. It is easy to understand why a new instrument dealer is almost bound to decline to let a brand new instrument out of his sight. Some don't seem too happy about a customer even picking one up. The person selling a used sextant, however: a) knows that the instrument is no longer pristine; b) no longer treasures it; c) knows he is lucky to have a prospective buyer; and d) recognizes the request as prudent. If the owner will not let you have it for an afternoon, give him your phone number and ask him to call if he changes his mind.

Once you have the instrument for a trial, in addition to running the vital checks, you should take some sights with it both to get the feel of the instrument and see if it will give consistent readings. A good sea horizon is best, but not essential for this test. The idea is to take sun sights at regular intervals, such as a minute, for a period of say 10 minutes before and after Local Apparent Noon (LAN) and to plot the altitudes opposite the times on a sheet of graph paper. In taking the sights, it is helpful to have an assistant to record the figures and give you a mark at the time to take the sight. Lacking a sea horizon, the ridge of a building roof to the south will do as long as it appears level from your aspect and covers a wide enough segment of horizon to serve during the westward movement of the sun.

Incidentally, a rough time of LAN for this test can be derived from the times of sunrise and sunset from the newspaper if *The Nautical Almanac* is not available. Half the time span between those times marks LAN when the sun is due south of your position in north latitudes. For example, if sunrise is 0730 and sunset at 1640, there are nine hours and 10 minutes between. Half of that—four hours and 35 minutes—added to 0730 puts LAN at 1205.

The roof ridge won't yield correct latitude as would the sea horizon, but only comparative figures are needed to test the performance of the instrument. If the plotted curve is reasonably smooth and fair, the instrument is performing well. If the plotted points fall irregularly, it may indicate excess lost motion in the instrument and an expert's opinion of the value of the instrument is in order. It still could be a good instrument and the irregularities caused by the user.

If not able to take a series of sights for LAN, a good test can be made with one sight of the sun with the sextant held horizontally to eliminate the distortions of refraction. The sun should be above 20 degrees. The reading in minutes and tenths of the arc from one side to the other can be checked against the semidiameter figure in *The Nautical Almanac* for the date. Appendix A gives an outline of the process.

A one-step check for lost motion is to set the instrument to zero and sight on the horizon or a distant roof line. A double image of the horizon line will appear unless there is no index error. Adjust the micrometer drum to make the two images coincide and note the reading. If it is off the arc, adjust the microdrum further in the same direction and then back in the other direction until the two images coincide again. The difference between the first and second reading of index error is due to lost motion. This procedure is referred to as coming on from each side. It gives an idea of the amount of slack in the gear system. If greater than a minute of arc, it is a problem that should be checked out by a technician. It may be correctable, or it may be from serious wear of gear teeth and thus a terminal condition.

As a note, coming on from different directions is ordinarily to be avoided when using the sextant for actual sights. One should try to come on to the horizon from the same side each time—and find index error coming on from the same direction.

This will minimize the effects from sight to sight of whatever lost motion is present.

If any of the foregoing raises doubt about the instrument's worth, do not hesitate to take it to a professional. Ordinarily, the cost for this evaluation is minimal—often it is free depending on the condition of the instrument and the mood of the instrument man. If the condition is so bad that he can tell at a glance you shouldn't buy it, he will be inclined to tell you so and not charge you. If it is a good instrument and you leave it with him for routine cleaning and adjustment, you are likely to get off without any charge for the opinion. In between there may be a charge if he had to spend some appreciable time making a determination.

On the subject of costs of repairs, it has been my experience that prices are remarkably moderate. Sextant dealers and repair people are an ethical group. If you go to a dealer who sends out his repair work, however, there will be an extra charge and extra time for sending it off. Better to take it yourself to the shop of the person who will do the work. Sixty dollars is a usual charge for cleaning, adjustment and lubrication. A paint job costs $125 to $150. The instrument must be disassembled for painting. I really don't like the idea of painting the whole instrument just for cosmetic reasons. As with a used car, painting has overtones of frame bending, collisions and cover-up. Better to leave signs of a little honest wear. Touch up nicks with instrument paint.

The single most frequent real repair required to rejuvenate a used instrument—that is, beyond simple adjustment—is mirror resilvering. It costs about $60, or $45 if the owner takes them off.

Remounting is not difficult, but a certain amount of manual dexterity is required in manipulating the spring clips that hold the mirror firmly, yet allow room for adjustment.

These stiff little springs must be stressed and held in order to get the holding screw back in its hole. It's a two-man job.

People tend to believe it is too risky to buy a used sextant because it may have been dropped and the owner is trying to unload it. It is possible to tell if the sextant has been dropped and seriously damaged by examining it closely and trying it out. Use a magnifying glass to check for any sign of a dent or flat point of impact. If there are none, and the sights you take with it are regular and consistent, you can feel confident it is sound.

The ultimate concern is does it take good sights? The best proof would be a round of sights under good conditions from a known position. This can be done in the back yard with an artificial horizon. These are effective on land and easy to set up at no cost. See Appendix F. Compared with the known position such a fix would give a definite indication of the accuracy of the instrument.

There is a way to buy a used sextant that provides both the low price and the assurance that the instrument is sound and true.

In fact, you get a certificate that guarantees it. The buyer can take the instrument on approval, use it long enough to get used to it, and return it for a refund if not satisfied. There is no way to beat an arrangement like that. To get a list of what is available, write or call Thomas Foulkes, 4B Sansom Road, Leytonstone, London E11, England (Phone: 01-539-5627) and you will receive a list of instruments in stock at that moment. All have been overhauled and recertified by an independent authority to determine their accuracy and general worthiness. A list I received in mid-1984 described a C. Plath, "As new," with a price equivalent of $348. Other prices ranged to as low as $150. Packing and Shipping to the U.S. was less than $40. There is no U.S. duty on sextants.

In response to a query about their policy on customers returning instruments, Mr. J.A. Cranmer assured me that Foulkes will take any instrument back and refund the full price regardless of why it was returned. If the customer simply doesn't fancy the instrument, he will have to pay the transportation costs. He added that no instrument has been returned in the 15 years the company has been selling used sextants. What could be more fair?

So, whether to buy a used sextant comes down to a matter of money. There is no reason to buy a used sextant if expense is no problem. It is not true that they used to be better made in the good old days. This is one item that today is made better than ever.

SELECTING A SEXTANT

There are about seventeen companies worldwide making sextants. Some of these are instrument companies that make sextants only when there is sufficient demand. The Germans and Japanese make the greatest share, by far, estimated as high as 95 percent with the British, American, Dutch, Germans, French and Chinese splitting the rest. Each company in this tiny industry makes a serviceable instrument. While other machines have increased in complexity as they were improved, sextants have not changed greatly over the years. Radical design changes have tended to be unsuccessful either technically or commercially.

The sextant is a relatively simple instrument and came to full flower with the introduction of the microdrum tangent screw and enlarged mirrors and horizon glass in the early 1900s. The companies that have survived and continue to make sextants are selling refinements in convenience and quality rather than innovation.

While some instruments are superior to others, all the sextant manufacturers make adequate instruments which exceed in accuracy the capabilities of most users under most conditions. This excludes lifeboat sextants and practice models.

Where differences among instruments are significant, however, is in the matter of whether they are made of brass, aluminum or plastic, their sizes, adjustability, repairability and cost.

I'll not duck the issue by saying they're all great. A respected professional nautical instrument maker friend says that the Tamaya full size instrument, the Jupiter, is easily the best value in a sextant. Its quality is close to that of German instruments, it is fully adjustable and repairable from the point of view of the instrument man, and it costs around 30 percent less than the comparable German instrument ($1,265 vs. $1,402). The Spica at the top of the Tamaya line is a fine instrument, but it costs about the same as its German equivalent and is thus not as good a value.

Tamaya, which is related to the Asahi Pentax camera company, is believed to be the largest maker of sextants. This is the estimate of one of the major U.S. importers based on sales—actual figures of production are trade secrets. That source also gave me an idea of the size and trend of the U.S. market. In the latter 1970s, when the sextant market was most active, about 2000 instruments a year were imported and sold. In 1984, sales were down by half. The most important reason for this decline has been the shrinking of the U.S. merchant marine in the wake of the reduction in exports due to the high priced U.S. dollar and resultant large fall in U.S. exports.

C. Plath is believed to be second only to Tamaya in current production. Traditionally, the German sextants have been widely considered to be the world's finest. C. (for Carl) Plath was the grand old man who first opened his instrument shop in Hamburg, in 1862. In the early 1960s C. Plath was acquired by Litton Industries, which needed a German firm to build its LN-3 inertial navigation system for the F-104G Starfighter. For a few years in the mid-1960s C. Plath was so busy building LN-3's that it subcontracted its sextant production. Litton never imposed American production methods on C. Plath's sextant

Fig. 5-1 C. Plath Navistar Classic. The trapezoid frame webbing and fat micrometer drum knob are distinctive. Photo courtesy of C. Plath.

production, however, and the company continues to produce quality instruments.

Dr. N. Kliemann, C. Plath Managing Director, has stated that the union with Litton Industries has had only beneficial effects in that C. Plath profits from the technology transfer from the other elements of

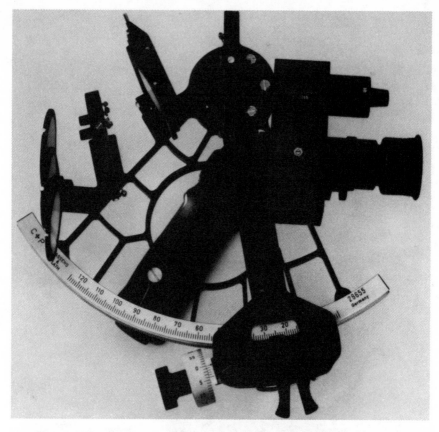

Fig. 5-2 Cassens & Plath with bubble attachment. Photo courtesy of Baker, Lyman & Co.

Litton, while maintaining its independence. Sextant manufacture is but a small part of the company's main production of ship control and guidance equipment, including some of the very electronic satellite navigation systems that now compete with sextants.

The other Plath is Cassens & Plath GMBH of Bremerhaven but there is no present connection between the two companies. In 1908, in Bremerhaven, the firm of Cassens and Bennecke lost Herr Bennecke by retirement and gained a new partner and considerable prestige when it was joined by Theodor Plath, son of Carl. Cassens & Plath made its own sextants, which, while remarkably similar to C. Plath instruments were and are distinguishable by a different interior frame design and several details.

Theodor personally was a partner with Cassens, but there was no intermarriage of the two firms and there is no connection today.

The sextants of the two companies look similar, but have several different features. The most obvious is the round centered frame of the Cassens & Plath and the rectangular center opening of the C. Plath frame. Another distinctive difference is in the design of the micrometer drums. It is interesting that in the Tamaya line, the most expensive model, the Spica, copies the C. Plath rectangular frame design while the Tamaya Jupiter has the round hole frame of the Cassens.

PHYSICAL CHARACTERISTICS OF CONCERN

Bowditch says, ". . .for accurate work . . . the radius of the arc should be six and a half inches or more." This warning dates from the days when the bigger the sextant was, the more accurately the arc could be divided because the degrees were bigger.

I don't know of any modern sextants being manufactured with a radius of greater than 6.48 inches. The yacht or 7/8 sized sextant is not only less accurate on account of its shortened pivot-to-arc distance, typically 5 1/2 inches, but is likely to have greater non-adjustable error—distortion of the machined arc. Plastic sextants have virtues such as very low relative cost, imperviousness to corrosion and a certain degree of toughness. In a test drop from about four feet high to a rug covered floor, a Carver plastic lifeboat sextant suffered a cracked index arm end. The bounce was about three inches and my pulse rate increased dramatically during the experiment. The chief utility of the plastic sextant in my view is for contingency back-up work and use on very bad days when you don't want to expose your number one instrument to the weather and probably won't be able to get highly accurate sights anyway.

The selection of one sextant in preference to another may come down, in the end, to the characteristics of the optics that come with an instrument.

The steadiness of the navigator is also a limiting factor in the degree of magnification of the sextant optics. The consideration is the same as for binoculars—in fact the 6×30 scope is half a pair of binoculars. The greater the scope's power, the steadier the

Fig. 5-3 Tamaya Jupiter has become one of the world's best selling sextants.

holder must be. The most popular binocular size is 7x50 because that is as powerful as can be held still. That's with both hands and even so ship's officers are often seen to steady their binoculars by resting elbows on railings. Anyone who has strained to read a distant buoy number with powerful glasses knows that frustrated feeling of being able to see the number and comprehend that it is big enough to read but not being quite able to do so because it is jittering around so much.

The practical upper limit of sextant magnification is six power. I recently had the opportunity to evaluate a sextant with an eight power monocular and was glad I didn't have to buy it to get some experience with it. It was difficult to hold it still enough for star sights.

Fig. 5-4 Venus, a 7/8 size yacht sextant, made by Tamaya.

The lower powered scope, which is commonly the standard scope on many sextants, is shaped like the cooling tower of a nuclear plant. Called a star scope, it usually is 4x40 power, sometimes 3x40 and even 2x40. Not only is its magnification less, its field of vision is considerably smaller than the monocular type scopes.

This can be a disadvantage when searching the sky to find a body in the first place. The instrument must be closer to right on before the body can be seen in the field of view. This may not seem important, but can make quite a difference, particularly in taking star sights rapidly and in finding Venus in daylight. It seems somewhat ironic that it is harder to find stars with a star scope, but it is true. The indisputable strong point of the star scope is that it is easier to hold still in adverse conditions. Some yacht navigators feel strongly about this characteristic and maintain that any scope stronger than four power is no good on a yacht.

Another consideration about the six power is that it is reputedly inferior for star sights over a dim or hazy horizon, which is the classic acid test for the navigator. Supposedly the higher power intensifies the starlight until the already dim horizon is further obscured by the dazzle. Maybe, but I can see the dim horizon better with the six power and I know I can see the star itself better. One caution—with the higher power, the star may seem to have some appreciable girth rather than be just a point of light.

Sirius is a case in point and the planets do have appreciable diameter. If so, the body must be split with the horizon rather than just touch the horizon as would be done in a sun lower limb sight.

In summary about optics, I'd say, that the 6x30 is the most useful scope for the widest range of applications. More powerful scopes are hard to hold still. Four by 40 and lower powers are easier for rough weather, but handicap you in average conditions.

The differences among the metal instruments are to be found in conveniences and price, snob appeal and fashion. What is the snob philosophy of sextants? The popular assumption used to be that the Germans made the best sextants, the Japanese copies of them were clever and usable, but hardly something one would want to be seen with, and the British instruments serviceable, but quaint—really suitable mainly for people who think cold showers before breakfast are character building. So the discriminating navigator might use a Japanese or British instrument out of curiosity, but would never buy one, for other than patriotic reasons. Other considerations aside, the non-German instruments had much less resale value.

In contemplating the comparison of features and prices of sextants, one thing is striking—the wide range of prices. One instrument is almost 10 times the price of another. Even ignoring the extremes and looking more toward the middle of the range, there are wide divergences between instruments of roughly equivalent capabilities. Not only do all these instruments accomplish the same single task, they all require the same

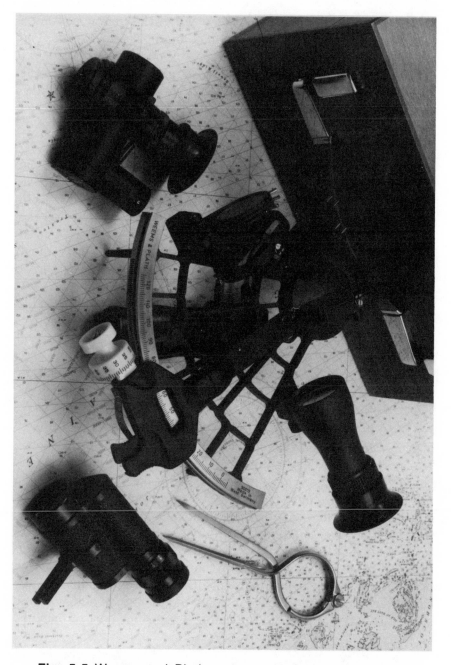

Fig. 5-5 Weems and Plath sextant with a 4x40 star scope mounted and a 6x30 monocular and a bubble attachment alongside. Photo courtesy of C. Plath.

series of operations, with the exception of the plastic sextants. They all are machine made to the maximum possible extent. The actual cost of production is inversely proportional to the retail price so that it should be possible to get greater discounts in the higher priced instruments. Sextants are like bathroom fixtures—priced high to allow multiple profit margins. So what you pay for an instrument may well depend on what level of the sales organization you deal with.

The very best way to select a new sextant is to go to the factory and buy it in the sales room. In 1980, after a year of comparative shopping, I bought a Cassens & Plath with 4x40 scope for $800. That was about $250 off the price tag and was the best deal I could find. Visiting Bremerhaven the next year I toured the factory and found I could have bought the same instrument for $500 there. Not too many months later, I could have bought it for about $400 because of the stronger dollar. In 1991, with the U.S. dollar weaker, the sextant prices have escalated in this country. When the U.S. price goes up, fewer are sold here and soon the surplus abroad tends to bring the price down. There used to be a 1.9 percent duty on imported sextants, but this has been eliminated.

Second to visiting the factory, the best way to buy certain makes of sextants is from a foreign mail order house such as Thomas Foulkes, of London.

Next best is dealing with the U.S. importers themselves, who may give you a discount price if the margin of profit is there. The mark-up between importer and small dealer is commonly 40 per cent. Sometimes these importers will take orders by phone, and give you a discount that way, but not always. Their problem is that you might be one of their dealers calling to check on whether they are being under-sold by their leader. They are apt to be somewhat reserved on the subject of discounts. The best thing to do is write a letter to the importer saying that you understand they give a 25 percent discount to merchant marine and other seamen and that you would like to buy model such and such, if that is the case. If you can manage some sort of marine-sounding address, so much the better. If it is in a town that doesn't have a dealer you are more likely to get a discount—perhaps as much as 20 per cent.

Walking into the importer's sales room gives you the potential of getting not only the discount, but an extra scope or other feature as well. Don't be intimidated by the fact that you may not be a professional seaman—if they get the discount, the importer can afford to give it to you too. If a sales person balks at your polite proposal, simply hand him your card with the request that if ever a sale should be held, please call you. And leave smiling discreetly. You may hear from him.

The most expensive way to buy a sextant is to walk into that little dealer in Port Sardine in a yachting cap and boat shoes with checkbook ready. Such a person represents the ideal customer. Even here it should be possible to get a discount in an over-stocked market. Remember that the dealer will make about 40 per cent and that he may not have sold a sextant lately. According to an official of a leading sextant importer, only about 1,500 were sold in the U.S. in 1991, and more than half of these were plastic.

CHAPTER 6

SIGHTING TECHNIQUES

It has been said of the porcupine that there is only one correct way to grab it—very carefully. The same is true with sextants, although there are two good ways to pick them up, both requiring care. It's a good thing there are two ways, because in its case, the sextant handle is underneath the frame and inaccessible. The fingers of the left hand, spread out like a clutching hand from the grave, are laced among the lattice work of the frame without touching index arm, scope or mirrors—the tender parts. Instruments are picked up in some dreadful ways—by the telescope, for instance. I once saw a TV drama about Robert Edwin Peary's discovery of the North Pole in which the great explorer was portrayed as actually tossing his sextant to an assistant after taking a sight! Holding the instrument by the frame also is necessary when handing it to another person; the handle is not big enough for two hands to grab. One of the first routines the Navy student navigator learns is the ritual for accepting responsibility when passing the instrument to another.

"Have you got it?" asks the passer.

"Yessir! I have it!" responds the receiver in a loud, firm voice, presumably so that if it then falls to the deck, witnesses can testify knowledgeably. I have never heard of a Navy sextant crashing to the deck accompanied by cries of, "I thought you had it!"

Actually, it is probably a better practice not to pass the instrument from person to person at all, but to put it down for the receiver to pick up himself. On a sailing yacht with no safe place to put it down, parking it on a sextant hook (Chapter 4) is a good solution.

Having picked up the instrument from the case with the left hand, next the handle is grasped with the right unless the scope must be attached first. If the case will permit it, it is best to leave the telescope in its mounted position, but some cases will not close unless the scope is removed. If the instrument has that V-grooved clamp instead of a collar mount, it may be safer to put the instrument down on its legs or on a sextant hook while tightening the clamp lug—a two handed task.

The natural tendency when sighting through a sextant scope is to close the left eye, but this is not the best practice for two reasons: fatigue and balance. It is hard to keep one eye closed for long periods and some effort must be expended to do so. It is far better to leave the left eye open and staring at the continuation of the horizon extending unmagnified off to the left.

Balance, however, is important when taking sights. The stance should be with feet comfortably spread an appropriate distance for counteracting whatever motion the ship may have. The upper body must be free to sway around while keeping the sextant on target.

Fig. 6-1 Sextant is inverted for initial acquisition of body and rough alignment with the horizon. Drawing courtesy of Defense Mapping Agency.

One surprise for many new navigators is how difficult it is to sight a high altitude body and then keep it in view while bringing it down to the horizon. The body tends to get lost during the process—even a body as big as the sun with its path of reflection on the water to help. There are two techniques to try here. The first is inverting the sextant during the search. To do this, first set the index at zero so that both the direct and reflected lines of sight are pointed in the same direction. Then turn the sextant upside down and get a rough alignment with the target by sighting along the outside of the scope. This is easy with a straight scope and a little more awkward with a prismatic scope with its offset after body. Then look through the scope and center it on the body. You've got it made now—just release the tangent screw clamp and glide the index arm forward. The horizon will appear to rise to the level of the body. As soon as the horizon is in the field of view with the body, turn the sextant upright and make the final adjustment with the microdrum in the usual way.

This technique is much easier than trying to bring the body down to the horizon, which requires coordinating the movement of the index arm along the arc with the rate of depression of the entire instrument, thus keeping the body continuously in view. The other method for finding the target is the use of the pre-calculated azimuth and altitude of the body as mentioned in Chapter 9. The index arm is set to the nearest whole degree of the pre-calculated altitude. Then the body is sighted over a compass on the prescribed azimuth. Most ships have a gyro compass repeater on each wing of the bridge and some on top of the pilot house as well. They read in degrees true, that is, angles measured clockwise from the direction of the geographical North Pole.

This is the same reference used in the tables—and *HO 214, 229* and *249*. These repeaters are mounted chest high and it is a simple matter to sight through the preset sextant over the flat dial along the azimuth where the body is scheduled to be. The results still repeatedly thrill and amaze me.

In most cases the desired body will be centered more or less in the middle of the field of vision of the scope. One has only to refine the alignment with the microdrum, and record the sight. This technique is particularly impressive when used for daytime sights of Venus. Venus is visible in daylight when its altitude is

Fig. 6-2 Gyro Sextant, Kreisel-sextanten KS-3D, made by C. Plath during World War II, is rare today. Powered by compressed air, it used an averaging device and took sights without reference to the horizon—the ultimate achievement for a sextant. Seventeen inches long, it was heavy and cumbersome, but far ahead of its time. Photo courtesy of G.D. Dunlap.

greater than that of the sun and not too close to it. The trouble is finding it.

Against a light blue sky, it is almost impossible to spot at random, even scanning with binoculars. Yet with pre-calculated altitude and azimuth, it will pop out in your sextant telescope clearly visible.

Yachtsmen seldom have a gyro compass to use with pre-calculated azimuths, but a magnetic compass will serve nearly as well provided it is the right design and conveniently situated, such as the kind mounted on a binacle, which often has a large easily read horizontal card that is illuminated and legible from all directions.

If a boat has only an edge reading compass, one can make a pelorus by cutting a compass rose out of an old chart, or better, out of a small area plotting sheet. The latter is easier to read. Mount this on a thin piece of plywood or anything suitable to give it weight and stiffness. Cover it with something transparent to keep spray off. Rubber feet on the bottom will keep it from slipping around. Put this on the seat or deck in front of the sighting position and align it with the ship's heading just before sighting over it. The altitudes of pre-planned stars can be written in grease pencil right on the pelorus for maximum efficiency.

One thing to be careful about with a magnetic compass is temporary deviation caused by a belt buckle as you belly up to the compass for sights, or perhaps by the sextant itself if it has ferrous metal parts or batteries in the handle. To determine whether there is a problem—observe the compass closely to see whether there is any change in reading as you approach from various angles with your gear in sighting position. Ideally, one would like to be able to hold the sextant a couple of feet above and behind the compass so that the navigator can easily look down and tell in which direction he is pointing the instrument.

Another important consideration in using pre-calculated azimuths is magnetic variation. It must be applied to the true degree values of Zn from the HO 249 tables or star finder in order to get the magnetic compass azimuth needed to find the body. A variation of 10 degrees, not allowed for, can prevent finding the body quickly and spoil the advantage of pre-planning.

One side result of the use of this locating technique is that modern navigators are not as constellation conscious as those of the past. The constellations served to help locate and identify stars so navigators paid attention to them. But the trouble with depending on constellations for navigation star identification is that they have either faded out already, or are lagging behind the navigational stars just when you want to refer to them.

Keeping in mind that most star sighting is confined to the twilight periods when there is a horizon from which to measure, and that the navigational star is usually the brightest in its part of the sky, it can be understood why modern navigators aren't much concerned with constellations. At evening twilight, the navigator must have completed his round of sights long before the gathering darkness reveals the rest of the constellations arrayed around the navigational stars.

Swinging an arc before measuring the altitude of a body is an important procedure, often only casually touched upon by sex-

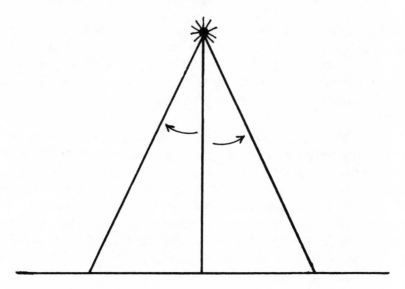

Fig. 6-3 Only the vertical distance to the horizon is accurate. Slant distances are too long.

tant instructions. Its purpose is to help in judging when the plane of the sextant frame is vertical, a condition that must exist if the sight is to be accurate. If the sextant is leaning to one side when the measurement is made, it will be measuring a diagonal distance greater than the true altitude of the body. See **Figure 6-3**.

The final line of position resulting after the sight is reduced to a line on the chart will be in error toward the body. Davis Instruments, which makes a prism device for helping to keep the sextant vertical (Chapter 3), has calculated that a 6 degree slant when sighting a star at an altitude of 45 degrees results in an approximate navigation error of 10 miles.

Swinging an arc is also called rocking the sextant and simply means rotating the instrument from side to side around the line of sight to the horizon.

As the index mirror slants from side to side the reflected image of the body being superimposed near the horizon will describe an arc, high on the ends and low in the middle. See **Figure 6-4**.

When the reflection of the body is at the low point of this arc, the sextant is vertical; when off toward the high ends of the arc, the sextant is cocked and a reading taken then would be in error.

To demonstrate the importance of taking the sight on the vertical, deliberately rock the sextant to one side, adjust the microdrum to match the body to the horizon and note the reading. Then compare that reading with one taken with the body at the bottom of the arc. The amount of error depends on the altitude of the body as well as the slant of the sextant, but is approximately one mile for each minute of difference in the two readings.

In swinging an arc, the left edge of the mirrored right side of the horizon glass is a big help. The body should be halved by this edge at the moment when the instrument is vertical and the body appears at the bottom of the arc, as depicted in the center panel of **Figure 6-4**. The tilted side panels of the figure show that when the sextant is rocked, the body that was tangent to the horizon appears to rise above it, transcribing an arc such as seen in the lower part of the sketch.

One problem in swinging an arc is that the plumb sextant attitude is gone almost as soon as perceived.

In order to be of any value, the swinging of the arc must be coordinated precisely with the actual sight. The sextant must be rocked to the right; then, while in the act of rocking back toward the left, adjust the microdrum so that the bottom of the arc will very nearly be tangent to the horizon. On completion of the swing to the left, swing back to the right and note if the body is tangent at the bottom of the arc and if not, which way to adjust to make it so. During perhaps the third swing with the body at the bottom of the curve and exactly bisected by the vertical edge, the final tweak is given to the microdrum to bring the body into perfect tangency and the time is marked. A good sight requires that the absolute minimum possible time lapse between determining that the instrument is in the right position and the taking of the sight.

A common mistake is swinging an arc that is centered a degree or so above the horizon and then cranking the body down to take the sight. No good. In anything other than ideal conditions, such as when becalmed on a lake, the sextant will have been moved out of the vertical in the interval. Some textbook illustrations with explanations of swinging an arc tend to foster this error by showing the arc of a body some distance above the horizon. The bottom of the curve of the arc must achieve tangency at the very instant of the sight, especially in rough weather, when it is most difficult.

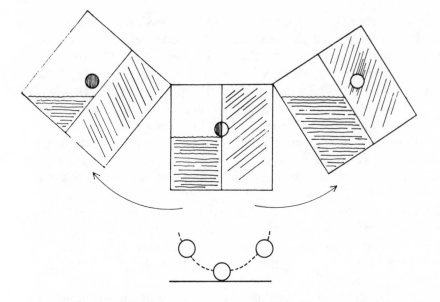

Fig. 6-4 Three views through the horizon glass of a sextant being rocked from side to side. Bottom illustration traces the apparent arc made by the body during the rocking.

One more wrinkle on swinging arcs. The technique works best on bodies at medium altitudes, say 30 to 45 degrees. With bodies at low altitudes, the arc flattens out and becomes such a slight curve that identifying its low point is tough.

As the sextant is rocked to one side, the body simply scoots off stage. At high altitudes, the arc becomes hard to manage because of its sharp curve. Swinging the arc is important enough to justify selecting only mid-altitude bodies, if there is any choice. It would not be wise, however, to decline to make an observation of any navigational body no matter how high or low when ship's position is in doubt.

SUN'S UPPER LIMB

Most navigators use the upper limb of the sun only when the lower limb is obscured. When the sun is at lower altitudes the longer diagonal look through the earth's atmosphere causes greater distortion the closer to the horizon. The refraction distortion of the upper limb is substantially less than that of the lower at altitudes of less than about 20 degrees. Over the years, I have found that I get better results using it all the time, unless it is obscured. I think part of the reason is that I like the upper limb better. In the morning, it is first to appear and substantiates my faith that the terrors of the night have been overcome once again. The upper limb is the last to leave in the evening and does so only reluctantly with a promise to return, evidenced occasionally by a green wink.

The practical benefit of using the upper limb stems from the superior contrast between the sun's incandescent disk and the background of the sea below the horizon. As I bring the sun up to the horizon, the gap of darker water narrows until the upper limb just kisses the sky at the horizon. Ah, the poetry of the thing!

Swinging the arc is a bit different for upper limb sights in that the arc crosses the horizon, but this causes no difficulty. I like the concept of using the upper limb at LAN when the point is, after all, to find the height of the sun's climb at our meridian.

The question of the effect of irradiation on the accuracy of a sun sight is controversial. Irradiation is the phenomenon causing the eye to perceive a light area or object against a darker background as larger than it actually is. It can affect the horizon, which it makes appear lower, and the sun, which it enlarges. These two effects tend to cancel each other out if the lower limb is being matched to the horizon, but to augment each other if the upper limb is being brought up to the horizon. The *Nautical Almanac* from 1958 to 1970 included a constant correction for irradiation for upper limb sights of the sun. Newer almanacs do not because, as Bowditch explains (Art 1609), the degree of the effect depends on the individual observer, the size of his scope, the altitude of the sun "and other variables." In other words, it is impossible to quantify. All this might seem to make it best for the navigator to avoid upper limb sun shots as many authorities do recommend, but I get consistently better results using the upper limb routinely in spite of irradiation.

I suggest, during some happy period when things are going well and there is reason to believe your position is accurate, that a few upper limb sun lines be taken for comparison. The best test is to find a site ashore with a good sea horizon to the south and take a series of sights alternating upper and lower limbs for latitude at noon. Comparing results with the charted latitude of the location will tell the story.

The probable cause of the blindness of Galileo, maker of telescopes and champion of heliocentricity, was too much gazing at the sun without sufficient protection. The sun is the single most important navigational body. There is little danger of damage to the eyes for two reasons: low powered telescopes and efficient shades. Galileo made and used magnifiers of up to 30 power while the highest power sextant scope currently in use is eight.

Most sextants come equipped with seven shades—disks of darkened glass of varied densities—usually a series of four for the index mirror and three for the horizon mirror. They are swung in or out of the line of sight as needed to shield the eye.

The sun is tamed down to a bright disk with a distinct rim and can be viewed indefinitely without discomfort. The usual practice is to start out sun sights using the darkest glass on the index mirror and the lightest on the horizon glass. Then the next lighter and the next darker respectively are tried to see if there is an improvement. Often the first arrangement turns out best.

If the sextant is being used in an inverted position for initial rough alignment of sun and horizon, the navigator must start with the darkest shade over the horizon glass and the lightest over the index mirror—the reverse of the upright arrangement. Otherwise a blast of too bright sunlight in the eyeball will result. Then the shades must be reversed when the sextant is returned to the upright position. This sounds like a lot of trouble and is indeed bothersome. A better system is to know the approximate altitude in advance and set it in the sextant. No tables are needed. One can judge the altitude of a body simply by using horizontal fingers held at arm's length as a guide. Four of my fingers held at arm's length are about 10 degrees wide. Just alternate hands up the sky to the body for an approximate altitude. Best to check your own measurements against known altitudes. I may have fat fingers or short arms.

Some sextants have variable density or color filters in the lines of sight of the index mirror and horizon glass. These can be

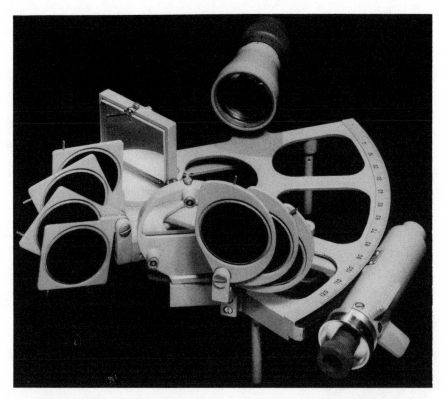

Fig. 6-5 Shade glasses fanned for display on a Zeiss/Freiberger instrument. The battery holder for arc illumination attaches to the side of the cylinder at the bottom of the index arm. Photo courtesy of Baker, Lyman & Co.

located in place of the shades or even inside the telescope housing as in the Dutch-made Observator sextant. Most filters work on the Polaroid principle and are adjusted by twisting an outer rim.

One thing about shades or filters that may not be fully appreciated is how helpful they sometimes can be after dark. The most obvious use is with a bright moon. When viewed by telescope, the moon can dazzle the eye. A shade over the index mirror can prevent this and make the edge of the limb more sharply defined. Sometimes moon sights can be taken well after twilight but the truthfulness of the horizon must be evaluated carefully.

Some navigators feel they can never trust a horizon illuminated by the moon—that it is an illusion.

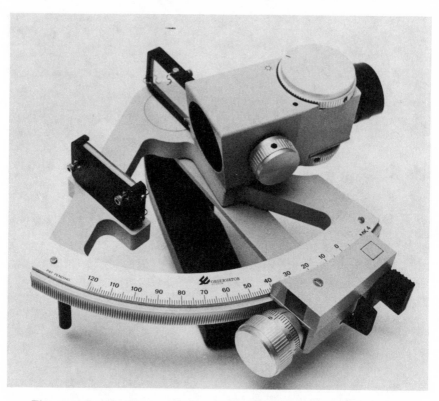

Fig. 6-6 Dutch Observator sextant. Shades for both the horizon and the reflected view are built into the monocular and are controlled with the large knobs on each side. Photo courtesy of Observator Instrument Engineers, Rotterdam.

My experience has been that it is often possible to tell whether the reflection of the moonlight on the water is distorting the horizon by studying it through a shade of light density over the horizon glass. Sweeping slowly from one side to the other across the sparkling water of the path of reflected moon light can reveal whether the horizon under the moon is in one straight line with the true horizon where they merge on either side of the path. Often it becomes apparent that what looked like a good horizon under the moon is not aligned with the real horizon and thus useless.

Or is it? How long has it been since the last fix?

It depends on how badly a navigator needs a line of position. With a landfall due soon and a lack of recent evidence of DR position validity, I suggest taking the moon sight over its suspect horizon using the darkest index mirror shade feasible.

Then set the sextant to its current index correction in order to make the reflected horizon coincide exactly with the directly viewed horizon.

Aimed now at the moonlit horizon with the moon excluded from the field of vision, and no shades on, it is often possible to distinguish a discontinuity between the segment of horizon under the moon and the true horizon off to the side fading into obscurity. The horizon line across the sparkling moonlight path on the water is likely to appear elevated above the true horizon because of the irradiation effect. The horizon should be examined also with the lightest shade over the horizon glass to see if that reveals a step up or down at the edges of the moon path. If a step is detected, adjust the microdrum to move the reflected horizon to the level of the true horizon. This yields the difference in degrees and minutes between the true and false horizons which can be applied to the sextant altitude like an index error correction. Plotting the sight with and without the false horizon correction will show which is most compatible with the DR position. After landfall is made, it might be worth the trouble to figure back to evaluate the correction in light of subsequent events.

Star altitudes can be measured by moonlight on some occasions depending on their position relative to the moon. A careful evaluation of the horizon is best made with the aid of the 7x50 binocular with its wide view and good resolution. Often it can be seen immediately that the apparent horizon is made only of moonbeams. On the other hand, binoculars can reveal that it is a perfect, sharp line joined exactly with the dark horizon and thus good for a round of sights. While doing this, it is best to avoid looking at the moon itself. I find it hard to overcome the temptation to take a quick look as long as I've got the binoculars out anyway—and always later regret the loss of dark adaptation. I know one prudent and solemn navigator who wears a green city room eye shade when the moon is bright, and he would never look at it through binoculars.

Light shades also are of some use with the brightest stars and planets—Venus and Sirius in particular. Unshaded these bright bodies, in some atmospheric conditions, can have an irregular glare or twinkle, like distant auto headlights.

Fig. 6-7 The horizon under the moon may appear elevated above the adjoining dark horizon, but is usable with a correction for the difference.

Absolutely clean optics will help to define the target better.

The lightest shade over the index mirror will help a great deal in taming Venus down so that her distracting glitter is eliminated. When stars or planets appear as more than a point of light, their mass should be split with the horizon. When I learned to appreciate this nicety I noticed an improvement in the sizes of my cocked hats. That term refers to the triangle formed at the intersections of the inaccurate Lines of Position of an imperfect three-body fix. It resembles a naval officer's tricorn hat. The smaller the hat, the better the fix.

ROUGH WEATHER TECHNIQUES

As the weather deteriorates and the sea builds toward conditions that finally become impossible for taking sights, the navigator has to adjust both his program and his techniques. Probably the first item to drop from the agenda as the weather worsens is star sights. These require a calm, deliberate approach which is difficult to achieve if the craft is corkscrewing around in a cross sea, however staunch one's philosophy may be. By lying to or running, sun lines can be had in fairly boisterous conditions.

The thing that helps in rough water is consistency of timing the sight. Call it rhythm. There usually is a repeating cycle in pitching, yawing and rolling that can be identified by paying attention to the signs and evaluating them. The most vexing problem is rolling and the resultant plunging and soaring of the target. Five or 10 minutes before attempting a sun sight, without the sextant in hand, the navigator should stand on the spot where the sights will be taken to see how she rides. Is it possible to stand there, hands at sides, swaying back and forth with the roll without clutching something? How about bracing stern-to against the bulkhead and leaning with the roll from the waist? On a brisk early spring day in the Atlantic aboard a Morgan 41, I found the best I could manage was sitting cross legged on deck with my back braced against the mast during rolls to starboard.

During rolls to port, where the sun was, I had enough freedom to level on target for a moment or two.

It is a great help actually to time the rolling cycle; there is usually some element of consistency revealed. While listening to the WWV time tick, watch an inclinometer pendulum as it travels from extreme to extreme. Note the number of beats from port to even keel to starboard. Count with the beat. Note whether every seventh swell is bigger, as old shellbacks avow. Bigger swells usually move the boat more sedately and are thus better for sighting.

Having analyzed the motion, get out the sextant and set it to the anticipated altitude of the body. Assume whatever stance you have found best and get in synchronization with the roll by counting. The object is to take several sights of the body in the same spot of the same phase of several rolls. The best time is as the roll away from the sun is just completed and there is a hesitation before commencing the roll back. On some ships, this halcyon interval can be quite substantial. On yachts, the roll will depend on any number of factors ranging from keel design to how much sail is being carried and how much liquid remains in tanks. The roll may be slow and graceful or snappy such as that Eric Hiscock complained about aboard *Wanderer III*. Aboard an improperly loaded freighter, it can be long enough to make drops of blood break out on the chief mate's brow. When the ship has rolled back away from the sun one will have to lean forward to stay on target and then match it to the horizon during the coming lull. This leaning forward is beneficial psychologically; it gives the navigator an upbeat feeling and inspires him. The alignment of the sun with the horizon should be made on the same limb and from the same direction on each of the succession of sights. If one roll feels noticeably different, discard the sight unless it is that big one recurring on schedule.

From a series of rough weather sights, the single sight that felt best could be used.

It is uncanny how, with experience, the navigator can tell by intuition which sights are most accurate. It is more prudent, however, to plot the whole series on graph paper.

The thought of graphing sights discourages many and it is rather a waste of time in good weather unless there are special circumstances. In rough weather, however, when the opportunity of getting good information is reduced and the probability of bad sights is increased, it makes sense to massage what information is available to put it in the best possible shape.

The graph need not be elaborate or detailed. Use any style paper with regular scales in both directions. The kind with plain quarter inch squares all over is good. Mark off convenient increments up the left edge of altitude spanning the value of the first and last sights taken. Across the bottom make a time scale that covers the interval. Plot the altitude of each sight taken above its time on the scale. If all of the series of sights were perfect, they would plot in a straight line (for a short interval) ascending or descending according to the time of day. Chances are good, however, that all the sights will be in error to some degree so that they will define a zig-zag pattern through which, with a transparent straight edge, a line can be drawn judiciously with some points above and some below. If all points are near or on the line, use a value in the middle of the series. If none are on or near the line, use an arbitrary point on the line at the middle time.

Thus the scatter effect experienced during a round of rough

weather sights can be averaged out. M.J. Rantzen, in *Little Ship Astro-Navigation*, points out that in the case of very erratic sights, graphing may not produce the best average because wild shots may coincidentally line up better than good ones. He described an ingenious method of using *HO 229* for establishing what the slope of the line of sights *should be* for the series. In a

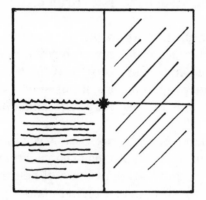

Fig. 6-8 Large stars and planets must be split by the horizon.

set of widely varying sights, any two of which are both on the slope line are valid. The advantage of graphing, with or without slope line, is that wild shots are usually easy to identify and eliminate, whereas in mathematical averaging, wild values are more likely to skew the average undetected. Either method, however, is superior to trusting a single uncorroborated rough weather sight.

Aboard a yacht, in addition to synchronizing sights to the motion, it is essential to take the sight from the very top of the marching mountain of water as it surges past. This not only gives the navigator the best view, but insures that the horizon is somewhere back out in its normal range for height of eye of the craft. It also tends to put the boat in the same attitude for successive sights. It is not necessary to try to estimate the actual wave height and correct sights for a different height of eye; the top of the wave which is distorting the horizon is presumed to be about the same height as the one being ridden by the yacht so that the difference should still be the normal height of eye.

And a good thing too, for which of us is audacious enough to think he can estimate wave height more accurately than little, big and gigantic. If you think you can judge feet by comparison with the mast think back to the last time you sweated your way under a bridge of uncertain clearance.

A 50-foot mast looks for all the world like it will scrape the grating on the Chesapeake Bay Bridge, which is 182 feet at center span.

About the only way I know for the navigator to approximate wave height at sea is to observe what portion of a distant approaching vessel's masts are obscured for you when you are on top of a wave and swell and then, if possible, ask the skipper how tall his mast is. This is a rough gauge and won't work at short range.

With irregular rolling, jerky cork-screwing motion, elevator ascents and ski jump descents, has the yacht navigator got much chance at all in rough weather? Some—if he can master what are referred to as snap sights. The term means sights quickly taken at the optimum moment. At the pinnacle of one wave, the sun is brought into rough alignment with the horizon. At the top of the next crest, tangency is touched up quickly and the time marked. Wild shots, which can be identified by an intensification of that sickly feeling lurking in the stomach, are discarded summarily. The good shots *feel* good.

EYEGLASSES—THE NAVIGATOR'S CURSE

Persons with 20/20 eyesight will only realize what a superb advantage they had in taking sights after they have lost it. Wearing glasses is a great disadvantage to the celestial navigator because of the incompatibility of seeing well through the sextant scope and reading figures on the arc in dim light interchangeably. One or the other can be easily done, but not both alternately.

The scope serves to gather more light from the objective and thus make it easier to see. Bodies with appreciable areas are magnified; starlight is intensified. The scope can compensate for poor eyesight of the routine variety. A person who needs thick glasses to read can adjust a telescope to see moon craters quite well without glasses.

Scope eyepieces sometimes are provided with a foldable rubber rim intended to accommodate users with or without glasses. Extended, it holds the naked eye at the optimum distance away from the rear viewing lens of the scope. Folded back it permits the eyeglass lens to be placed closer to the rear scope lens where it works better. This style folding eyepiece is helpful for eyeglass wearers—another advantage is that the folded rubber rim is suitable to rest the eyeglass lens against for steadying purposes. A non-folding rubber eye cup that is shaped something like a spittoon is not as satisfactory for eyeglass use because one can't get his eye as close to the ocular lens as intended and thus the field of view is depressingly reduced.

Less satisfactory still for the eyeglass wearer is the plain hard-rimmed telescope lens against which the eyeglass lens must be pressed to see properly. The eyeglass lens may be scratched and tends to slip around making the sighting difficult and depriving one of the appreciable steadying effect of a telescope securely pressed against the head. Least satisfactory is the shaped rubber eyepiece made to fit the face contours around the eye. Being asymmetrical, it is almost impossible to position an eyeglass lens against so that the view is comfortable and centered. But it is fine for viewing without eyeglasses and gives the best face support of all, putting the head to work effectively helping to steady the instrument.

What many people end up doing is taking the sight bare eyed with the scope adjusted to compensate for vision defects and then putting on the glasses for reading time and altitude.

Bifocal eyeglasses, otherwise such a boon, are especially troublesome in sight taking.

The line between the two different lenses, which does not intrude in ordinary use, is conspicuous and in the way when looking through a sextant scope.

It tends to distort the horizon line or the body or both. The navigator must be sure that he is looking at both body and horizon through the same strength lens by holding his head tilted just right.

I was discussing this matter with another navigator one time and he jokingly suggested that what was required is a bifocal monocle. The left eye could do the close work through a lens,

Fig. 6-9 Author models homemade navigator's monocle which permits bare-eyed sighting with one eye and corrected vision scale reading with the other. Photo by Nancy Bauer.

leaving the right eye unencumbered for the telescope. Not a bad idea. So I made myself a monocle with ear pieces by removing the right lens of an old pair of glasses and cutting off the bottom of the frame to permit access for the telescope eyepiece.

This device works pretty well though it does have a couple of minor drawbacks.

Casual looking around with them on is distracting at first, and there is a strong tendency to close one eye or the other. Then there are a lot of funny remarks from bystanders—witty stuff about standing too close to the sextant.

I also have seen navigators wearing granny glasses, those lower-half, semi-circular lenses. If one has rather a long nose, they can be worn down far enough so that the sextant eyepiece can be aligned above the right lens and the glasses remain in place for reading. This solution might be good for those whose distant vision remains sharp.

The last word on eyeglasses is that they are extra troublesome for sextant work but can be used so as to minimize their drawbacks. If one can't see well without them, contact lenses may offer a viable alternative.

CHAPTER 7

CORRECTING
THE SIGHT

The movement of heavenly bodies is precise. Try as he may, man is not. That's why there are corrections to be made when dealing with celestial navigation—as many as 13. Some are for fairly sophisticated reasons. Others, such as *personal error* (PE), are debatable. To factor this error in, the error must be consistent and I don't think one can be *consistently* wrong any more than one can eliminate sighting errors entirely. Let's be reasonable—sometimes you get pinpoint fixes and sometimes you don't.

Still, this and the other dozen errors need to be dealt with, and sometimes for seemingly inexplicable reasons. For instance, it has always seemed rather odd that the value determined by observing a body is not the *observed altitude* (Ho). It is only *sextant altitude* (Hs) and must be refined with other data before it is fit enough to be called Ho and compared with *computed altitude* (Hc). There is even an intermediate state called *apparent altitude* (Ha), reached after three corrections have been applied to the raw sight.

The first correction to be considered, *instrument error* (I), compensates for a combination of the three built-in errors that cannot be adjusted out of a sextant.

These are: prismatic errors of the shades, eccentricity of the limb, and graduation errors. Their magnitude depends directly on the quality of the instrument.

In years gone by, when sextants were hand made, these er-

C↔P

CASSENS & PLATH
2850 Bremerhaven, Germany

SEXTANT - ATTEST / CERTIFICATE

Nr. 29992 / No.

Nr. 18881 / No.

Buch 15 / Book

Hersteller: Cassens & Plath
Manufactured by:

Trommelsextant / Micrometer Sextant

Ablesung Reading	Teilung Graduation	Beschaffenheit Condition	Optik Optics	Beschaffenheit Condition
Gradbogen Arc	1°	gut/good	Femrohr Telescope	gut/good
Nonius Vernier	0.2'	gut/good	Spiegel Mirrors	gut/good
Trommel Micrometer	1'	gut/good	Vdkl.-Gläser Shades	gut/good

Vergrößerung / Magnification: 4 x 40

Halbmesser des Gradbogens / Radius of Arc: 160 ___ mm

Anzubringende Verbesserungen wegen Exzentrizität des Gerätes
Corrections because of eccentricity

bei/at	0°	10°	20°	30°	40°	50°	60°	70°	80°	90°	100°	110°	120°
Berichtigung Correction													

Dieses Instrument ist für den Gebrauch als fehlerfrei zu bezeichnen.
This instrument is free of errors for practical use.

CASSENS & PLATH

Bremerhaven, 07.03.1980

Fig. 7-1 Cassens & Plath certificate. On the back is a statement that the instrument error is no more than plus/minus 9 seconds of arc or .15 minutes. Although the graduation of the micrometer is 0.2 minutes of arc, the instrument can be read simply by reading between the lines.

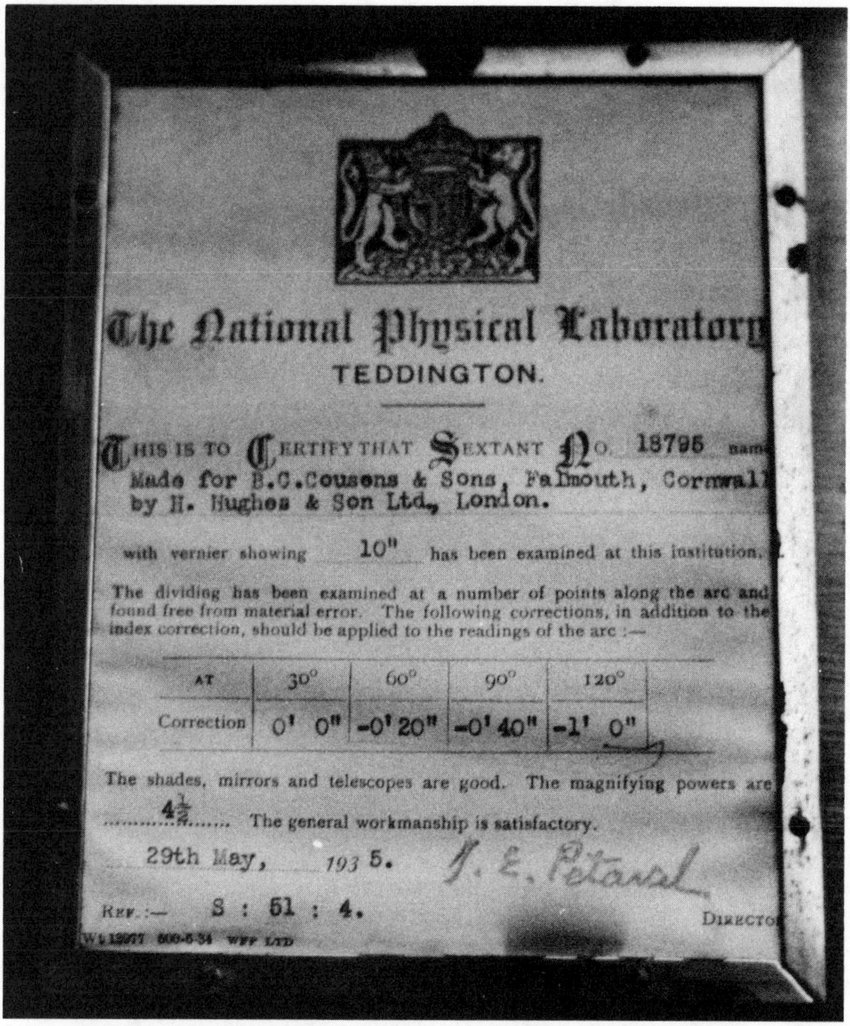

The National Physical Laboratory

TEDDINGTON.

This is to Certify that Sextant No. 18795 nam[e]
Made for B.C.Cousens & Sons, Falmouth, Cornwall
by H. Hughes & Son Ltd., London.

with vernier showing 10" has been examined at this institution.

The dividing has been examined at a number of points along the arc and found free from material error. The following corrections, in addition to the index correction, should be applied to the readings of the arc :—

AT	30°	60°	90°	120°
Correction	0' 0"	-0' 20"	-0' 40"	-1' 0"

The shades, mirrors and telescopes are good. The magnifying powers are 4½ .... The general workmanship is satisfactory.

29th May, 1935.

Ref.:— S : 51 : 4.

Director

Fig. 7-2 Certificate of testing of a Hughes sextant issued by the British National Physical Laboratory from which the instrument corrections are taken when correcting a sight. Interpolation is necessary; the correction for a sight of 45° would be –0′10″. Courtesy of Col. Warren Davis.

rors were a regular instrument feature. Today, with computer controlled production techniques, most modern sextants are more than adequately accurate. In fact Cassens & Plath will not ship any instrument that is out more than plus or minus 9 seconds of arc, which it considers to be substantially error-free.

This isn't true of all makers, though, and the custom of affixing a certificate of error on the lid inside the sextant case remains. Most manufacturers conduct tests and fill in the certifi-

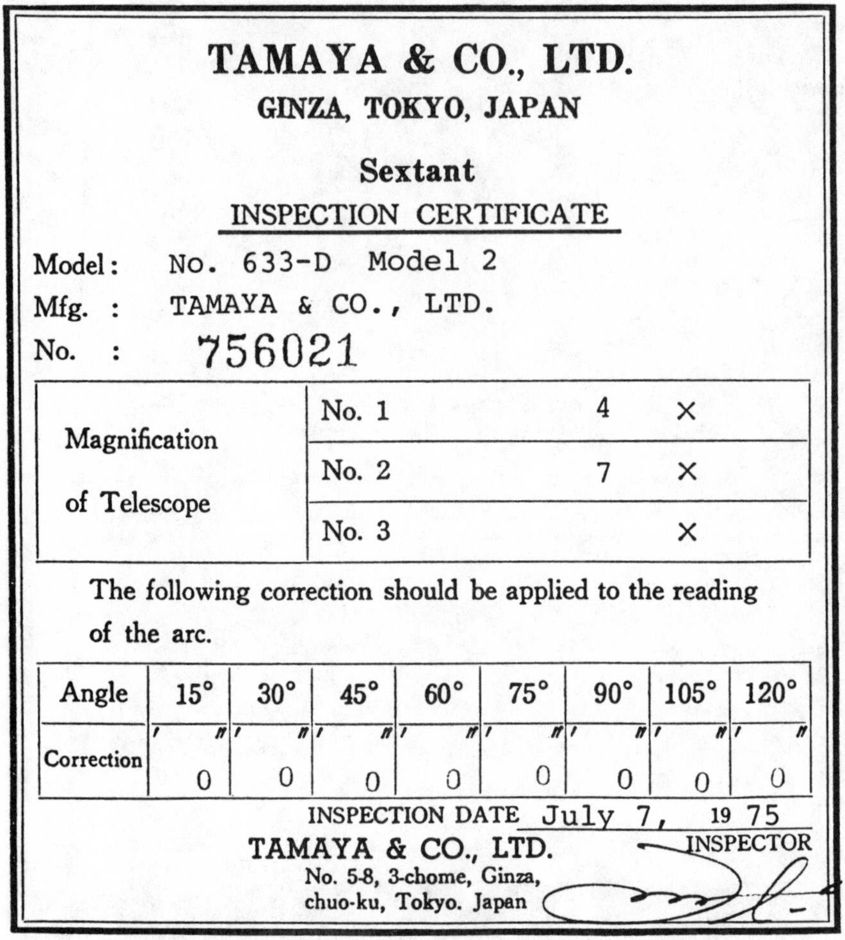

TAMAYA & CO., LTD.
GINZA, TOKYO, JAPAN

Sextant
INSPECTION CERTIFICATE

Model :   No. 633-D Model 2

Mfg. :   TAMAYA & CO., LTD.

No. :   756021

Magnification of Telescope	No. 1	4	×
	No. 2	7	×
	No. 3		×

The following correction should be applied to the reading of the arc.

Angle	15°	30°	45°	60°	75°	90°	105°	120°
Correction	0	0	0	0	0	0	0	0

INSPECTION DATE July 7, 19 75

TAMAYA & CO., LTD.   INSPECTOR
No. 5-8, 3-chome, Ginza,
chuo-ku, Tokyo. Japan

Fig. 7-3 Japanese certificate of inspection indicates a perfect instrument.

cates themselves, while the British have a system providing for independent evaluation whereby, for a fee, a sextant is submitted for testing to the National Physical Laboratory, at Teddington. The certificate has a table showing instrument error at intervals, usually 10 degrees, from one end of the arc to the other. See **Figure 7-2**.

So the instrument error taken from the certificate in the sextant box top is applied with reversed sign to sextant altitude. Select the value nearest the altitude of the sight, or interpolate if midway between two significantly different values.

The values are usually given in seconds of arc and will need conversion (divide by six) to tenths of minutes of arc.

The second correction to be applied to the sextant altitude is the *index correction* (IC), which is the equivalent of index error with sign reversed.

It is caused by the index mirror and horizon glass not being exactly parallel when the sextant is set at zero.

Index errors that are on the arc, that is positive, are simply subtracted out.

Remember, "If it's on, take it off."

Index errors off the arc, or in the "arc of excess" as the British say, can be misread easily. These values are minus readings from zero down.

On the micrometer drum sextant, it is not hard to appreciate that a reading of 58.4 minutes off the arc must be taken as minus 1.6 minutes of arc. It helps me to be sure which way from the zero reading the index error lies if I note the value and then move the drum back to zero and out to that reading again.

This leaves no doubt whether I have moved on or off the arc. In practice, index errors seem to vary back and forth a tenth of a minute of arc or two from day to day, but they usually stay on the same side of zero—unless the sextant has been adjusted right down to zero. Some navigators prefer to have a positive index error of one minute so that they always will have an index correction to apply in the same direction. This reduces the chance of applying a correction backwards and doubling the effect of the nonparallelism rather than eliminating it.

The third significant correction on the way to apparent altitude is for *dip* (D), commonly referred to as *height of eye correction*.

This is a very important correction, intended to compensate for the difference between a truly horizontal plane extending outward from the navigator's eyeball and the actual plane from eyeball to horizon which is tilted downward.

A simple way of thinking of this concept is to picture a person standing on the surface of a frozen sea. His feet are on a plane called *geoidal*, his eyes are in a plane with the *sensible horizon*. The downward angle to which the plane of sight must be adjusted to make the two horizons coincide is called dip or dip angle.

The size of the angle depends directly on the height of the eyeball above the surface, and to a lesser but significant extent, on how greatly light rays are being bent by atmospheric conditions, that is, refracted.

The dip table inside the front and back covers of *The Nautical Almanac* is constructed for standard atmospheric conditions but must be consulted first, whatever the conditions. The table is entered with height of eye, either in meters or feet. This means the distance between the surface of a calm sea and the eye of the navigator looking through the sextant from his position of observation. Inspection of the table will show that substantial differences in corrections occur for modest changes in height of eye in the low ranges. For example, the correction for 2 feet is 1.4 minutes of arc and for 4 feet, 1.9 minutes of arc. A minute of arc is the equivalent to a mile in plotting the fix, so being two feet off in height of eye can cause a half mile error. Therefore, it is worthwhile for a yachtsman *to measure* his height of eye from his customary observation spot rather than estimate. This is best accomplished by an assistant from the dock some quiet day alongside a pier.

The table of dip is described as a "critical table," which means that no interpolation is required, according to the explanation on Page 259 in the back of *The Nautical Almanac*. Although I'm uncomfortable criticizing the great book, that is only half true; the right hand column is noncritical and should be interpolated for best accuracy. Evidence of this can be deduced by noting that the arguments are on the same line with the corrections. In contrast, on the critical side of the table, a zone of heights of eye all get the same correction, written adjacent and midway between the limiting numbers. What to do at the border—the critical number? Read the upper correction. This refers to position on the page, not numerical value. The correction for 11 meters is -5.8 minutes of arc, not -5.9. It should be firmly fixed in the inner most recess of the very soul of all navigators that dip correction is always minus, even if the navigator is working from the lower end of a periscope in a submerged submarine. Those are not dashes in front of all the corrections in the table. They are minus signs, every one.

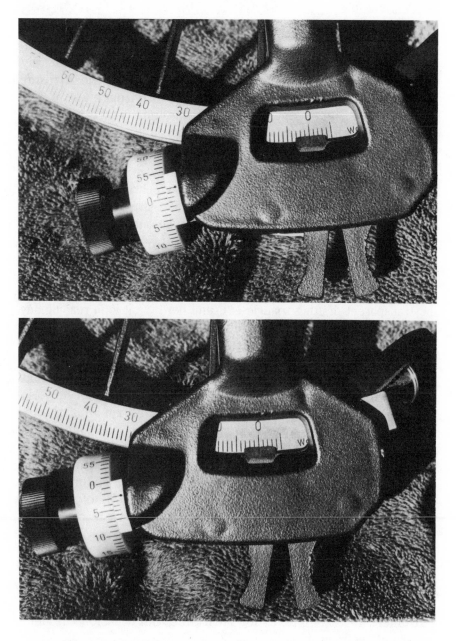

Fig. 7-4 Micrometer drum index error readings. Upper setting is *off the arc* 1.6 minutes of arc (60' – 58.4' = 1.6'). Lower view is 2.8' *on the arc*. Note that on the main arc scale, it is difficult to tell which way the index line has moved.

Having modified Hs into Ha with the application of three corrections, Ha is subject to further correction, both for standard conditions and for non-standard, if they exist. The best way to appreciate non-standard conditions is to study the *Altitude Correction Tables—Additional Corrections* on page A4 of *The Nautical Almanac.* Entered with Ha, temperature and barometric pressure the table produces a correction in minutes of arc and tenths with a plus or minus sign to be applied to Ha.

The table is initially awkward to use with its diagonal zone arrangement at the top half. It is easier to locate the intersection of the temperature and pressure values if a transparent drafting triangle is aligned with the two values so that the right angle corner falls in the proper diagonal zone. Having determined the zone, one goes to the bottom half of the table to a column of the same letter-name and reads off a correction and algebraic sign from a rank of values in line with the apparent altitude. The straightedge of the transparent drafting triangle is useful here, as it is in reading any table in *The Nautical Almanac.* Many mistakes can be avoided. Interpolation is advisable when there is more than a tenth of a minute of arc difference between corrections bracketing an apparent altitude.

Taking a broad view of the table for a moment, one notes that the zone labeled G contains zero value corrections for every Ha from 0 degrees to 50 degrees altitude—the full range of the table. Backing out of the table, it can be seen that the zone where no corrections are required goes from 68 degrees Farenheit at the higher barometric pressures, down to temperatures of 32 degrees Farenheit at the lower pressures. Thus is delimited the zone of standard conditions outside of which the Ha of a body is apt to be in error unless the table is used and the correction applied. The data center of the table is 50 degrees Farenheit temperature and 29.83 inches of pressure—the spot being marked by the cross below the G on the table. Notice how small a change in either temperature or pressure will shift a sight into a zone of non-standard conditions. A change from 50 to 60 degrees Farenheit, for example, at low altitudes, can occasion the need for sizable corrections.

A final point about this often ignored table—the big corrections are found at the extremes.

A4 ALTITUDE CORRECTION TABLES—ADDITIONAL CORRECTIONS

ADDITIONAL REFRACTION CORRECTIONS FOR NON-STANDARD CONDITIONS

Temperature

−20 F. − 10   0°   +10°   20°   30°   40°   50°   60°   70°   80°   90°   100°F.

−30 C.   − 20   −10°   0°   +10°   20°   30°   40° C.

Pressure in millibars: 1050, 1030, 1010, 990, 970

Pressure in inches: 31·0, 30·5, 30·0, 29·5, 29·0

Zone letters: A B C D E F G H J K L M N

App. Alt.	A	B	C	D	E	F	G	H	J	K	L	M	N	App. Alt.
0 00	−6·9	−5·7	−4·6	−3·4	−2·3	−1·1	0·0	+1·1	+2·3	+3·4	+4·6	+5·7	+6·9	0 00
0 30	5·2	4·4	3·5	2·6	1·7	0·9	0·0	0·9	1·7	2·6	3·5	4·4	5·2	0 30
1 00	4·3	3·5	2·8	2·1	1·4	0·7	0·0	0·7	1·4	2·1	2·8	3·5	4·3	1 00
1 30	3·5	2·9	2·4	1·8	1·2	0·6	0·0	0·6	1·2	1·8	2·4	2·9	3·5	1 30
2 00	3·0	2·5	2·0	1·5	1·0	0·5	0·0	0·5	1·0	1·5	2·0	2·5	3·0	2 00
2 30	−2·5	−2·1	−1·6	−1·2	−0·8	−0·4	0·0	+0·4	+0·8	+1·2	+1·6	+2·1	−2·5	2 30
3 00	2·2	1·8	1·5	1·1	0·7	0·4	0·0	0·4	0·7	1·1	1·5	1·8	2·2	3 00
3 30	2·0	1·6	1·3	1·0	0·7	0·3	0·0	0·3	0·7	1·0	1·3	1·6	2·0	3 30
4 00	1·8	1·5	1·2	0·9	0·6	0·3	0·0	0·3	0·6	0·9	1·2	1·5	1·8	4 00
4 30	1·6	1·4	1·1	0·8	0·5	0·3	0·0	0·3	0·5	0·8	1·1	1·4	1·6	4 30
5 00	−1·5	−1·3	−1·0	−0·8	−0·5	−0·2	0·0	−0·2	+0·5	−0·8	+1·0	+1·3	+1·5	5 00
6	1·3	1·1	0·9	0·6	0·4	0·2	0·0	0·2	0·4	0·6	0·9	1·1	1·3	6
7	1·1	0·9	0·7	0·6	0·4	0·2	0·0	0·2	0·4	0·6	0·7	0·9	1·1	7
8	1·0	0·8	0·7	0·5	0·3	0·2	0·0	0·2	0·3	0·5	0·7	0·8	1·0	8
9	0·9	0·7	0·6	0·4	0·3	0·1	0·0	0·1	0·3	0·4	0·6	0·7	0·9	9
10 00	−0·8	−0·7	−0·5	−0·4	−0·3	−0·1	0·0	+0·1	+0·3	+0·4	−0·5	+0·7	+0·8	10 00
12	0·7	0·6	0·5	0·3	0·2	0·1	0·0	0·1	0·2	0·3	0·5	0·6	0·7	12
14	0·6	0·5	0·4	0·3	0·2	0·1	0·0	0·1	0·2	0·3	0·4	0·5	0·6	14
16	0·5	0·4	0·3	0·3	0·2	0·1	0·0	0·1	0·2	0·3	0·3	0·4	0·5	16
18	0·4	0·4	0·3	0·2	0·2	0·1	0·0	0·1	0·2	0·2	0·3	0·4	0·4	18
20 00	−0·4	−0·3	−0·3	−0·2	−0·1	−0·1	0·0	−0·1	+0·1	+0·2	+0·3	+0·3	+0·4	20 00
25	0·3	0·3	0·2	0·2	0·1	−0·1	0·0	+0·1	0·1	0·2	0·2	0·3	0·3	25
30	0·3	0·2	0·2	0·1	0·1	0·0	0·0	0·0	0·1	0·1	0·2	0·2	0·3	30
35	0·2	0·2	0·1	0·1	0·1	0·0	0·0	0·0	0·1	0·1	0·1	0·2	0·2	35
40	0·2	0·1	0·1	0·1	−0·1	0·0	0·0	0·0	+0·1	0·1	0·1	0·1	0·2	40
50 00	−0·1	−0·1	−0·1	−0·1	0·0	0·0	0·0	0·0	0·0	+0·1	−0·1	+0·1	+0·1	50 00

The graph is entered with arguments temperature and pressure to find a zone letter; using as arguments this zone letter and apparent altitude (sextant altitude corrected for dip), a correction is taken from the table. This correction is to be applied to the sextant altitude in addition to the corrections for standard conditions (for the Sun, stars and planets from page A2 and for the Moon from pages xxxiv and xxxv).

Fig. 7-5 Table A4 *The Nautical Almanac*. The diamond below G is the data center. The dashed lines are the boundaries of the zone of standard conditions. If temperature or pressure happen to be outside the zone, the table should be checked for applicability.

At low altitudes when the line of sight goes for a greater distance through the atmosphere, bigger corrections are needed. At extremes of temperature and pressure light rays are badly bent.

As a reminder to check this table for applicability, the careful navigator may want to note the temperature and pressure at the top of his worksheet along with the date and times of twilight and other key data.

The corrections for standard refraction conditions are found on the inside cover and next page of *The Nautical Almanac*. The main table, *Altitude Correction Tables 10—90 Degrees—Sun, Stars, Planets* is entered with apparent altitude. These tables, like the adjacent dip table are arranged as critical tables—the user is not to interpolate.

This is indicated by the way in which the correction is placed opposite the interval between two apparent altitudes not level with either value. Critical tables are much faster to use than those requiring interpolation. The other important feature of this table is its arrangement—by halves of the year for the sun and planets, and by upper and lower limbs of the sun. Selecting the wrong column, particularly the sun's limbs can really warp a fix. This is such a common error that it is a good place to start looking when trouble-shooting a bad fix.

If the body sighted was at an apparent altitude of 10 degrees or less, the A3 table is to be used instead of A2. It is a closer look at the low altitudes where the effect of refraction is most severe. Failing to make these corrections will result in a very skewed position. Look at the size of the corrections specified in the table. Consider that when first the lower rim of the setting sun appears to touch the horizon the sun is actually already entirely below the horizon. What is seen is its image refracted up and over. Many navigators never use this table because they avoid taking sights of a body that low. While bodies at higher altitudes are to be preferred, the low altitude corrections really do their work as designed and there is no good reason to reject low altitude bodies categorically. They can be very helpful in special circumstances such as the opportunity for a daytime combination of moon, Venus and sun.

Speaking of the moon, the main corrections for its apparent

ALTITUDE CORRECTIONS

SUN/STARS/PLANETS

Sextant Altitude .... Hs   _____

Instrument Correction ....+/-   _____

Index Correction ....+/- IC   _____

    Pg. A2 DIP .... -HE   _____

Apparent Altitude .... HA   _____

Refraction Correction
    Pgs. A2/A3/A4 .... +/- R_____

Observed Altitude .... Ho   _____

MOON

Apparent Altitude .... Ha   _____

    Pgs. xxxiv, xxxv
        Moon Table, upper .... +_____
        Moon Table, lower .... +_____

Upper Limb Obs. only ....-30_____

Observed Altitude .... Ho   _____

VENUS/MARS

Apparent Altitude .... Ha   _____

Refraction Correction
    Pgs. A2/A3/A4 .... +/- R_____

Additional Correction
    Pg. A2 (Parallax) ....   + _____

Observed Altitude .... Ho   _____

Fig. 7-6 Guide for correcting Sextant Altitude (Hs) to obtain Observed Altitude (Ho). Page numbers refer to *The Nautical Almanac*. Feel free to photocopy this guide.

altitude are not shown on A2 and A3, but are pulled from *Altitude Correction Tables—Moon* in the very back of *The Nautical Almanac*. These are always added to the apparent altitude. If the upper limb of the moon is observed, 30 minutes of arc must be subtracted from the corrected apparent altitude total to achieve the observed altitude. The moon, by the way is one of the most useful celestials for the navigator. The prejudice against its use is apparently based on its being considered too fast-moving to yield good results and the extra corrections cumbersome. Neither is true and the moon makes a particularly excellent aid when available in daylight for use with a sun line.

There is an additional correction for Venus or Mars to compensate for parallax. Changes in the altitude of the apparent center of Venus, which shifts with the change in its phases, were incorporated into the main data tabulations of GHA for Venus starting in 1985.

Under practical circumstances, how are all these corrections applied? Instrument error, index correction and dip are combined mentally and applied to sextant altitude to produce apparent altitude which is used to enter the yellow page tables for the refraction correction. Apparent altitude, plus or minus altitude correction, yields observed altitude. That's it, unless you have noticed that temperature or pressure is non-standard by some sign such as icicles on the spreaders or a distortion of the solar orb near the horizon. Then you go for additional altitude correction tables on A4. And visit the back of the Almanac when you have measured the moon.

Some navigators go astray in applying one of these corrections—dip.

The table is based on a flat sea when the line of the horizon is quiet and straight and the angle of depression is constant, but what of the times when the man in the cockpit must alternately look upward to see the crest of the nearest swell which constitutes his horizon at the moment and then downward as his craft is hoisted aloft itself? And the man on the 70-foot high bridge of the loaded tanker plowing along through the same sea with so little motion that a nickel will stay balanced on its rim on the chart table—should they both apply dip correction as usual? In a neat bit of navigational irony, the yachtsman, very aware of the

Fig. 7-7 Correction for Wave Height. The dotted line represents the calm weather water level and horizon. The ship on left should use a correction of half the wave height in rough weather, but the yacht should not.

waves, tends to make big allowances when he should make none, while the smug mate on the big ship degrades his fixes sadly by not believing he must modify dip correction for wave height.

Why? The worlds of both boat and ship are really quite small—something like three or four miles in diameter from the cockpit and 10 or 12 miles from the bridge. The local wave system is likely to be uniform; the wave at the gunwale is about the same size as the one disturbing the horizon.

They were molded by the same forces. If the navigator in the ascending boat will but take care to take his sight when at the crest, his horizon will have been raised the same amount by a matching wave there and ordinary calm weather dip correction remains valid. The man high on the bridge of the smooth riding ship, on the other hand, must take into account that the difference between his eyeball height and the wave-distorted horizon is varying and that to average things out, he should be subtracting one half the estimated wave height from his height of eye when entering the dip table. The wave pattern superimposed on top of the swell will increase the height of the horizon, but will not raise the ship an equivalent amount as it would a yacht.

CHAPTER 8

TIMING THE
SIGHT ACCURATELY

Time is the indispensable element of celestial navigation, while *The Nautical Almanac* is a second-by-second tabulation of the angular positions of the navigational bodies throughout the year. If the navigator doesn't know exactly when an observation was made, he cannot use the timetable properly.

The actual timing is done with a comparing watch, jauntily known as a hack watch. The hack is the base time taken from the ship's chronometer, which in years past, was a costly and delicate piece of equipment that demanded almost continuous attention. These days, the easy availability of accurate time by radio and inexpensive electronic watches is a blessing. Nearly any short wave radio will pick up international time broadcasts and many brands are internally powered by batteries. Some radios are manufactured to pick up only time signals.

Called time cubes, these are dedicated receivers pre-tuned to WWV and WWVH that fill the time requirement nicely. The range of these stations varies with atmospheric conditions, but enough of the world is reached by them to justify investing in a pre-tuned radio. It usually has a telescoping antenna on one side, but weak signals can be enhanced considerably with the use of a wire run up the backstay with a topping lift. Anyone undertaking a global voyage should have a traditional short wave receiver that can pick up the many other time broadcasts throughout the world as well.

It is not a new idea to team up a time cube and a battery

TIME

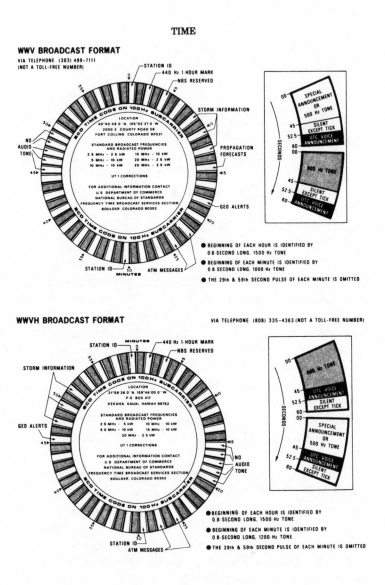

Fig. 8-1 Time Broadcast formats of stations WWV and WWVH. Courtesy of Defense Mapping Agency.

powered tape recorder to record the times and altitudes of celestial sights.

It is, however, only recently that the sizes and prices of the necessary machinery have become compact enough to motivate trying it out. With the reduction in size and the increase in battery life, the rig can be expected to run several months on a set of batteries. The technique eliminates the need for writing anything down while taking a sight.

The tape recorder is arranged next to the radio in such a way that the time signal will be recorded. This requires a recorder with a built-in microphone to record the radio.

It also must have a microphone jack of the type that accommodates a mike with an off/on button so that the action can be stopped and started from the hand-held microphone. There are a variety of recorders around which have these features.

In the case of a recorder that will use only one mike or the other, a radio shop can perform a simple wire crossing operation so both mikes will work simultaneously. The usual situation is that both mikes work, with the internal one at a reduced level as long as the external mike is plugged in. This is fine, since it is preferable for the time signal to be in the background.

The time tick is turned on and the recorder activated but stopped by the off button attached to the mike. Satisfy yourself that all is ready by seeing if the tape reels turn when the button is turned on. Read in the date, time of day and general observations such as sea state as part of this check. Later one can replay, replot and relive the celestial details of the voyage. After the body has been found and brought to horizon approximately, and with the time tick approaching an even minute, start the recorder. Adjust the sextant for final alignment and sometime after hearing the even minute tone, when ready, say, "Mark." Then read off the sextant altitude and, for good measure, record the azimuth and give an evaluation of the quality of the horizon, cloud cover or other condition that might later help you evaluate an inconsistent sight.

Stop the recorder and go to the next sight. The chief disadvantage is if the tape recorder jams up, data for a whole round of sights could be lost and not noticed until too late. A side benefit

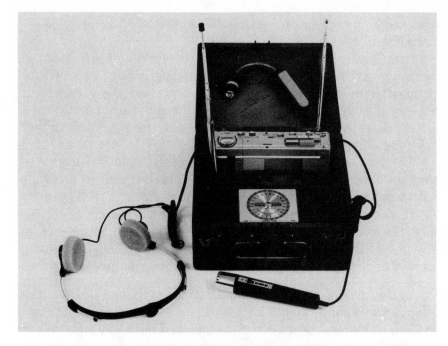

Fig. 8-2 Time Tick Tape Recorder. Broadcast shortwave time signals are picked up by the receiver in front of case and recorded with mark at time of sight superimposed by microphone on same tape. Photo by Scofield & Wolfe.

is the ability to record the intermittent weather reports and notices to mariners interspersed from time to time between minutes on WWV and which are sometimes audible when other broadcasts are not.

Before a round of sights, establish a routine to check and note the watch error along with the time from your wristwatch. This can save a lot of grief. A set of times could be reconstructed should the hack watch fail.

An ordinary stopwatch makes an acceptable hack watch. Selecting some even increment of Greenwich Mean Time (GMT), the assistant starts the hack watch and records that reference. The watch is left running throughout the round of sights and the time increments are read off in reverse order, the fastest changing increments first—that is seconds, minutes and hours.

An easier timepiece to use is called a split-hand stopwatch or stopwatch with lap timer. With these watches, the time of a sight can be marked by pressing a button. This allows the indicated value to be read at leisure while the watch is simultaneously maintaining the original time. The stopped second hand or digital display will catch up with the uninterrupted time when the button is pressed again. Split timers are usually top grade stopwatches and often expensive.

A wristwatch with a sweep second hand makes a perfectly acceptable hack watch as long as it has been compared with the true time very recently to determine its error and won't stop on you.

Fig. 8-3 Split second hand stopwatch. While one hand maintains total time elapsed, the other hand stops for individual sights and then catches up with the first hand. Photo courtesy of C. Plath.

Most navigators, denied the luxury of assistance, must muddle through alone and there are many different techniques, some fairly quaint. One method formerly common in the merchant marine is counting paces. It's used mainly by the old style up-through-the-hawse-pipe mates who are pretty phlegmatic about navigation in general and sextants in particular. The practice consists of taking a sight and then plodding methodically back into the chart house muttering, "One thousand one, one thousand two," etc., step by step. On arrival, the chronometer in its glass covered well at one end of the chart table is observed and the time written on the work sheet used for reducing the sight. The number of one-second paces is subtracted from chronometer time when chronometer error is applied. If the mate stumbles enroute or somebody asks him a question and interrupts his train of thought he starts over again with another sight.

For the yachtsman, standing in the companionway to take the sight and then ducking below to record it, counting seconds during the interval is reasonably accurate. The interval is short enough that any error is likely to be small. If the WWV time tick is left on and audible a very accurate time can be kept simply by noting the number of ticks heard in the interval between taking the sight and reading the time.

A much better way to time a sight is to wear a quartz wrist-watch with both an analogue dial and a digital display. The digital is set to keep GMT and compared with a short wave signal daily to establish its rate of change or watch error in case some day the radio signal cannot be received. I have tried a number of ways of recording the latest error on the watch itself, rather than in a log, which is just one more thing to keep track of.

The modern low-priced quartz watch is remarkably accurate, with some gaining only about two tenths of a second per day in a perfectly regular fashion. Batteries for them typically last more than a year, however, I put a new one in every New Year's day as a sort of commitment to better fixes in the coming year and for insurance against a dead battery at a critical time.

At sight time, I transfer the watch to the inside of my right wrist and start the stopwatch in synchronization with a convenient hack time. There, the watch can be found without looking and the lap/stop button pressed by the left hand in what I estimate to be exactly one second. If I should fumble for the button, I make it two seconds. One can get a good indication of the time involved by repeating the motion several times with the time tick broadcast going in the background. I mentally subtract the time correction as I write down the sextant altitude so that I won't have to bother with it later.

A stopwatch is not absolutely necessary for timing sights.

I know an extraordinary yacht navigator, Alex Fowler, of Annapolis, Maryland, who fastens an ordinary watch with a sweep second hand to the index arm of the sextant. Having taken a sight, he turns the sextant over calmly and reads first the watch and then the arc, both with aplomb and accuracy. He also allows a second for the interval and gets splendid results with no fuss at all.

Probably the most common means of self timing is the use of a sweep second hand pocket watch held in the left hand. The cord through the bale at the top of the watch is wrapped around the hand so that the watch cannot slip away. Having taken the sight, it is easy to look at the watch in the left hand, however the procedure of transferring the sextant to the left hand so the results can be recorded is complicated. Fewer fingers are free to grasp the frame and the grip is more awkward and less secure. One solution is to have a longish cord around your neck for holding the watch after it has been read.

It is not a new idea to juryrig a stopwatch to the handle of a sextant in as convenient a manner as possible so that its button can be pressed with the thumb at the instant the sight is taken. This eliminates any interval between sight and reading the time. I have seen watches taped to sextant handles in various ways and talked to navigators who were enthusiastic about the advantages. One common drawback is that the sextant cannot be put back in its case with the watch attached.

There is one British company, Vega Instruments, that sells a replacement handle with a stopwatch built into its upper end. The handle fits the East German Freiberger sextant made by Zeiss. I wrote to Vega and asked whether they made handles to fit other sextants and was told no because most other sextants have a battery compartment in the handle and no space for a watch. The Freiberger's battery compartment is attached to the end of the index arm and the standard handle is solid wood. Vega mentioned the possibility of adapting a Timex stopwatch to other types of sextants–a project they were contemplating. I bought a Timex Lap Stopwatch for about $23 and experimented with various ways of attachment. It is water resistant, light, compact and best of all, the activating buttons are large and they give a little squeak when pressed to let you know the message was received. It makes a good sight timer and is easily attached to many styles of sextant handles. See **Figure 8-4**.

The wrinkle that makes the idea workable is positioning the timer on the inside of the handle at the top end. There are multiple advantages to having it there rather than outside. First, the lap timing button faces forward and is convenient to press with the index finger, which is the better digit for the job. The thumb is already employed holding the handle and better left alone. Second, the face of the timer can be read through the frame of the instrument from the same side as the arc. No awkward twisting of the instrument to read the time is required. The face of the timer is upright, which is not the case if it is attached on the outside of the handle. The handle will fit into its nest in the box with the timer mounted—but the swing latch, if such is provided, cannot be closed. Sponge pads glued to the lid over the index and horizon glass positions will hold the instrument in position just as well. When arduous transportation is anticipated, I remove the timer and reactivate the handle latch, among other precautions.

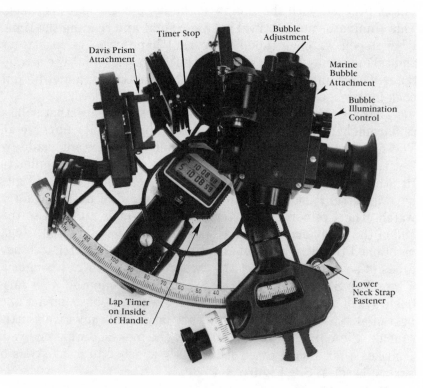

Timer Stop

Davis Prism Attachment

Bubble Adjustment

Marine Bubble Attachment

Bubble Illumination Control

Lower Neck Strap Fastener

Lap Timer on Inside of Handle

Fig. 8-4 A Cassens & Plath with Timex lap timer mounted on inside handle. The timer stop button is conveniently located for pressing with forefinger without disturbing grip or balance at critical instant. Photo by Scofield & Wolfe.

A good way to attach the timer to the handle is with nylon wire ties, those little nylon straps sold at electronic shops and used to bundle wiring together. Each tie has a ratchet and catch arrangement so that when the tie is drawn up snugly it stays put. Once tightly cinched, these ties cannot be loosened and must be cut to be removed. They're cheap and handy around the boat for

Fig. 8-5 Heuer Microsplit lap timer is small enough for attachment to the inside of the sextant handle like the Timex in Fig. 8-4. Its buttons are also big enough for easy activation. Courtesy of C. Plath.

other jobs, so I carry a wide assortment in my ditty bag whenever I put to sea.

Another timer, the Heuer Microsplit 1000, sells for about $45 and is only slightly larger than the Timex. It keeps regular time as well as being a lap timer, has large, easy-to-activate buttons, and is suitable for strapping to the sextant handle like the Timex.

Cassens & Plath offers a timer for $125 that comes with a bracket to attach it to the top of the frame between it and the handle.

It is thumb-activated and read through the frame. They are not ordinarily stocked by dealers and must be ordered from Germany. Persons considering ordering one should appreciate that the timer is actually a wristwatch. The control buttons are tiny and must be pushed very carefully or they won't function. They are decidedly inferior to the big, hearty buttons of either the Timex or the Heuer.

CHAPTER 9

SEARCHING FOR STARS

One of the most common causes for poor results in celestial navigation is haste brought on by ill preparation.

The mate on the morning watch who starts to think about taking sights only when he happens to notice that it is getting light, is too late to do anything but scamper after a fast fading star or two.

The taking of a sun line in full daylight is less time critical, but deliberate advance preparation is necessary. Certainly if one hasn't figured the approximate time of local apparent noon (LAN) in advance he either is going to have to do a lot of unnecessary waiting around or risk missing the sun's transit of the meridian.

The first step is to determine when the action is going to take place.

What time is sunrise? From the daily page of *The Nautical Almanac* that time as well as those of the twilight—nautical and civil are taken—and written down in a particular spot reserved for them such as the upper right hand corner of a dated page in a steno pad.

On average, there is only about a 20-minute prime period for star sights and none of it should be wasted.

The navigator next will figure which stars and planets will be in favorable positions and precisely where to look for them. There is a popular misconception that, in action, the navigator scans back and forth across the sky until he spies a star he knows, or alternately simply selects one at random. In practice, the conscientious person figures out and writes down the predicted positions of half a dozen or more of the best and most favorably arrayed bodies.

Ideally, the bodies selected should be spaced around the 360-degree celestial panorama at approximately equal intervals with no very acute angles between any two. Neither is it good to select stars that are directly opposite each other. There are several ways to find this information, but the two I like best are the device called the Rude Star Finder and a book, *HO Publication 249 - Sight Reduction Tables for Air Navigation.*

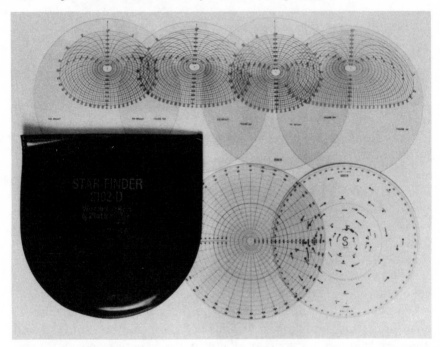

Fig. 9-1 Star Finder kit consists of one planet and eight star templates, a base plate, and a black vinyl pouch labeled 2102-D, which is believed to be a government euphemism for Rude, the name of the inventor. Courtesy of Defense Mapping Agency.

Capt. Gilbert T. Rude, United States Coast and Geodetic Survey, patented and later sold to the U.S. Government, an adjustable plastic chart of the heavens that can be preset using the local hour angle (LHA) of Aries. The Greenwich hour angle (GHA) taken from *The Nautical Almanac,* must be converted to LHA by formula. Turn to the first page of the Explanation section in the back of the book and in paragraph 2, titled *Principle,* is written the most basic equation in all celestial navigation:

$$\text{LHA} = \text{GHA} \begin{array}{c} \text{- west} \\ \text{+ east} \end{array} \text{longitude}$$

For any instant of time aboard ship, the time at Greenwich (GMT) can be determined by applying the zone description of the ship's position—one hour for each 15 degrees of longitude separating the two. The Almanac daily page is entered with the whole hours of GMT to find part of the GHA of Aries; the increment corresponding to the minutes of GMT is taken from the yellow Increments and Corrections pages in the back of the Almanac. The LHA of Aries at the ship is derived from this GHA of Aries by applying the DR longitude of the ship, subtracting if West.

Here is an example of the process for 0615, January 1, 1986 for a ship at 75° W:

Ship's time	0615	1 JAN 86	
Zone descrip.	+ 5		(75° W ÷ 15)
GMT	1115		

GHA Aries

	11 hrs	265°48.1'	(from NA daily pg)
	15 min	3°45.6'	(from NA yellow pg ix)
		269°33.7'	

$$\text{LHA} = \text{GHA} \begin{array}{c} \text{+East Long.} \\ \text{-West Long.} \end{array}$$ (from NA pg 254)

LHA Aries

	GHA Aries	269°33.7'	(subtract West.
	Long.	−75°W	If GHA too
		194°33.7	small, add 360°)

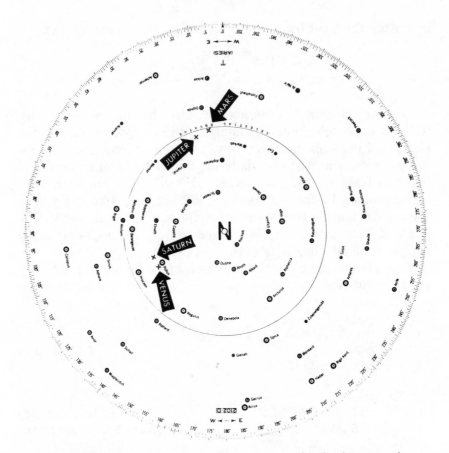

Fig. 9-2 A template for the nearest value of latitude, opposite page, is selected and placed over the base plate (above) with the template index arrow pointing at the proper LHA on the perimeter of the base. Azimuths and altitudes can be read off to within about five degrees for the principle navigational stars. Planets are plotted separately using another template. Courtesy of Defense Mapping Agency.

In figuring LHA of Aries for star-finding purposes, it is not necessary to be precise. All that is needed is the nearest whole degree. So the minutes can be rounded off (to 195 degrees in the example) without hurting the results.

The principal navigational stars are inscribed on the white base plate of the Star Finder. A transparent overlaying template, selected according to latitude, and oriented by LHA, enables the user to read off the bearing and altitude of the stars and planets which will be visible.

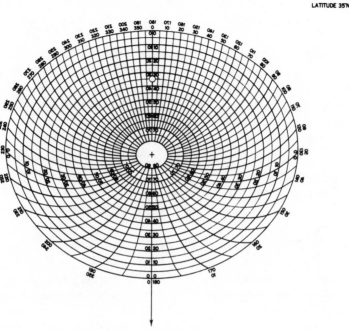

From the available choices, select and write down altitude and azi-muth of those high enough (above 15 degrees), low enough (below 65 degrees) and more or less evenly distributed around the entire sky. Sights can be taken of other stars, of course, but those at convenient altitudes are prime targets. Some stars are lucky for some navigators. I will go out of my way to get Altair, the flying eagle, or Vega, the fall-ing vulture; there is, after all, more to celestial work than mere skill.

When using the Rude Star Finder, some navigators like to secure the template in its proper setting with masking tape so it can be carried around without getting out of adjustment. A grease pencil arrow drawn to represent the course of the ship on the face of the template, makes a handy reference for orientating the stars in their positions. The preset Star Finder can be partic-ularly useful at times when the stars originally selected turn out to be obscured and alternates must be used. I sometimes carry the pre-set Star Finder around inside my shirt front when there are clouds that are likely to force me to alternates. I also make a written list of the targets in order by azimuths.

The list is quicker and easier to read in dim light than the Star Finder itself.

A word of caution on the subject of this dandy Star Finder— it melts easily and is useless thereafter. It comes in a nice black vinyl case which sucks up BTUs from sunlight like a sponge does water, particularly if left in some sheltered hot spot such as behind a wind screen.

Another handy way to get ready for the stars is using *Volume I, HO 249* which was designed as a simplified table for use by aviators.

That means, casting no aspersions skyward, that it was designed for fast, less precise navigation of the kind that airplanes require.

Air navigators must hurry or they are only recording history. These tables are a quick source of the position of the seven best stars in the sky at any given time and place.

They are selected for you—and are interchanged as time and latitude vary.

Volume I, HO 249 is entered with the latitude of the expected position of the sighting and the LHA of Aries for the expected time. Turning to the page marked with the appropriate latitude, the value of LHA is found in the left hand column and the selected seven, very best navigational stars are listed opposite. They are located in terms of Hc and Zn. Hc is calculated (or predicted) altitude of the body at the time in question; Zn is its azimuth or true bearing.

This must be adjusted for variation to yield a magnetic compass bearing. If the predicted altitude is set on the sextant and it is aimed in the direction indicated by Zn, the body will appear in sight almost as if by magic. This technique is most important in taking star sights successfully. It makes taking sights possible in substantially rougher conditions. If the sextant already is set to the right general value that awkward process of bringing the body down is eliminated. You can look for the body through the sextant directly.

The pre-planning information should be noted on something that is portable and easy to write on in the dark. The ultimate luxury is an illuminated clip board held by your assistant stand-

LAT 37°N

LHA ♈	♦CAPELLA Hc Zn	ALDEBARAN Hc Zn	♦Diphda Hc Zn	FOMALHAUT Hc Zn	ALTAIR Hc Zn	♦VEGA Hc Zn	Kochab Hc Zn	LHA ♈	♦Dubhe Hc Zn
0	32 43 054	26 40 089	34 00 168	21 38 195	27 08 260	28 21 299	24 49 348	90	38 37 036
1	33 22 054	27 28 090	34 09 169	21 25 196	26 21 261	27 39 299	24 39 349	91	39 05 036
2	34 01 055	28 16 090	34 18 170	21 12 197	25 33 262	26 58 300	24 30 349	92	39 33 036
3	34 40 055	29 04 091	34 26 171	20 58 197	24 46 262	26 16 300	24 21 349	93	40 02 036
4	35 20 055	29 52 091	34 33 172	20 43 198	23 58 263	25 35 301	24 12 349	94	40 30 036
5	35 59 056	30 40 092	34 39 173	20 28 199	23 11 263	24 54 301	24 03 350	95	40 58 036
6	36 39 056	31 28 093	34 44 175	20 12 200	22 23 264	24 13 301	23 54 350	96	41 26 036
7	37 18 056	32 16 093	34 48 176	19 55 201	21 35 265	23 32 302	23 46 350	97	41 55 036
8	37 58 056	33 04 094	34 51 177	19 37 202	20 48 265	22 51 302	23 38 350	98	42 23 036
9	38 38 057	33 51 094	34 53 178	19 19 203	20 00 266	22 11 303	23 30 351	99	42 51 036
10	39 18 057	34 39 095	34 54 179	19 00 204	19 12 267	21 31 303	23 22 351	100	43 19 036
11	39 58 057	35 27 096	34 54 180	18 41 204	18 24 267	20 50 304	23 14 351	101	43 48 036
12	40 39 057	36 14 096	34 53 182	18 21 205	17 36 268	20 11 304	23 07 351	102	44 16 036
13	41 19 058	37 02 097	34 52 183	18 00 206	16 49 268	19 31 304	23 00 352	103	44 44 036
14	42 00 058	37 50 098	34 49 184	17 38 207	16 01 269	18 52 305	22 53 352	104	45 13 036

LHA ♈	♦CAPELLA Hc Zn	BETELGEUSE Hc Zn	RIGEL Hc Zn	♦Diphda Hc Zn	Enif Hc Zn	♦DENEB Hc Zn	Kochab Hc Zn	LHA ♈	Kochab Hc Zn
15	42 40 058	17 35 094	15 32 113	34 45 185	38 04 251	41 48 301	22 46 352	105	28 35 016
16	43 21 058	18 23 095	16 16 114	34 40 186	37 18 252	41 07 301	22 40 352	106	28 48 016
17	44 02 058	19 11 095	17 00 115	34 35 187	36 33 253	40 26 302	22 34 353	107	29 01 016
18	44 42 059	19 59 096	17 43 115	34 28 188	35 47 254	39 46 302	22 28 353	108	29 15 016
19	45 23 059	20 46 097	18 27 116	34 21 190	35 01 254	39 05 302	22 22 353	109	29 28 017
20	46 04 059	21 34 097	19 09 117	34 12 191	34 14 255	38 24 302	22 16 353	110	29 42 017
21	46 46 059	22 21 098	19 52 118	34 03 192	33 28 256	37 44 303	22 11 354	111	29 56 017
22	47 27 059	23 09 098	20 34 118	33 52 193	32 41 257	37 04 303	22 06 354	112	30 10 017
23	48 08 059	23 56 099	21 16 119	33 41 194	31 55 257	36 24 303	22 01 354	113	30 24 017
24	48 49 060	24 43 100	21 58 120	33 29 195	31 08 258	35 44 304	21 57 355	114	30 39 017
25	49 31 060	25 31 100	22 40 121	33 16 196	30 21 259	35 04 304	21 52 355	115	30 53 018
26	50 12 060	26 18 101	23 21 121	33 02 197	29 34 259	34 24 304	21 48 355	116	31 07 018
27	50 53 060	27 05 102	24 01 122	32 47 199	28 47 260	33 45 304	21 44 355	117	31 22 018
28	51 35 060	27 51 102	24 42 123	32 31 200	27 59 261	33 05 305	21 40 356	118	31 37 018
29	52 17 060	28 38 103	25 22 124	32 15 201	27 12 261	32 26 305	21 37 356	119	31 52 018

LHA ♈	♦CAPELLA Hc Zn	BETELGEUSE Hc Zn	RIGEL Hc Zn	♦Diphda Hc Zn	Alpheratz Hc Zn	♦DENEB Hc Zn	Kochab Hc Zn	LHA ♈	♦Kochab Hc Zn
30	52 58 060	29 25 104	26 01 125	31 57 202	65 10 259	31 47 305	21 34 356	120	32 07 018
31	53 40 060	30 11 105	26 41 125	31 39 203	64 23 260	31 08 306	21 31 357	121	32 22 018
32	54 21 060	30 58 105	27 20 126	31 20 204	63 35 261	30 29 306	21 28 357	122	32 37 018
33	55 03 060	31 44 106	27 58 127	31 00 205	62 48 262	29 50 306	21 26 357	123	32 52 019
34	55 45 060	32 30 107	28 36 128	30 40 206	62 00 263	29 12 307	21 23 357	124	33 07 019
35	56 26 060	33 15 108	29 14 129	30 18 207	61 13 264	28 33 307	21 21 358	125	33 23 019
36	57 08 060	34 01 108	29 51 130	29 56 208	60 25 264	27 55 307	21 20 358	126	33 38 019
37	57 50 060	34 46 109	30 27 131	29 33 209	59 37 265	27 17 308	21 18 358	127	33 54 019
38	58 31 060	35 32 110	31 03 132	29 10 210	58 50 266	26 39 308	21 17 359	128	34 09 019
39	59 13 060	36 17 111	31 39 132	28 46 211	58 02 266	26 01 308	21 16 359	129	34 25 019
40	59 55 060	37 01 111	32 14 133	28 21 212	57 14 267	25 24 309	21 15 359	130	34 41 019
41	60 36 060	37 46 112	32 49 134	27 55 213	56 26 268	24 47 309	21 15 000	131	34 57 019
42	61 18 060	38 30 113	33 23 135	27 28 214	55 38 268	24 10 309	21 14 000	132	35 13 019
43	61 59 060	39 14 114	33 56 136	27 01 215	54 50 269	23 33 310	21 14 000	133	35 29 019
44	62 41 060	39 57 115	34 29 137	26 34 216	54 02 270	22 56 310	21 14 000	134	35 45 020

LHA ♈	♦Dubhe Hc Zn	POLLUX Hc Zn	SIRIUS Hc Zn	♦RIGEL Hc Zn	Diphda Hc Zn	♦Alpheratz Hc Zn	DENEB Hc Zn	LHA ♈	Kochab Hc Zn
45	19 48 026	30 49 076	14 44 125	35 01 138	26 05 217	53 15 270	22 20 311	135	36 01 020
46	20 09 026	31 36 077	15 23 125	35 33 139	25 36 218	52 27 271	21 43 311	136	36 17 020
47	20 30 026	32 23 077	16 02 126	36 04 140	25 07 218	51 39 271	21 07 311	137	36 33 020
48	20 51 027	33 09 078	16 40 127	36 34 141	24 37 219	50 51 272	20 31 312	138	36 49 020
49	21 13 027	33 56 078	17 18 128	37 03 143	24 06 220	50 03 273	19 56 312	139	37 05 020
50	21 35 027	34 43 079	17 56 128	37 32 144	23 35 221	49 15 273	19 20 313	140	37 22 020
51	21 57 028	35 30 079	18 33 129	38 00 145	23 03 222	48 27 274	18 45 313	141	37 38 020
52	22 19 028	36 17 080	19 10 130	38 27 146	22 31 223	47 39 274	18 10 313	142	37 54 020
53	22 41 028	37 05 080	19 47 131	38 54 147	21 58 224	46 52 275	17 36 314	143	38 10 020
54	23 04 028	37 52 081	20 23 132	39 20 148	21 25 225	46 04 275	17 01 314	144	38 27 020

Fig. 9-3 *Volume I, H.O. 249, Sight Reduction Tables for Air Navigation* provides a fast and simple means of preselecting suitable stars. Knowing LHA of Aries and latitude, predicted altitude (Hc) and Azimuth (Zn) are read out directly.

ing behind you ready to write. This is Navy style where there is plenty of hired help. On yachts, there may or may not be someone to help out. In the merchant marine, there is never any help—it would be overtime.

I have tried several systems, none of which is entirely satisfactory. I have carried a small pad in my shirt pocket and a penlight. This is acceptable for reference when looking for the star, but requires both hands when writing results. What to do with the sextant then is a problem. Many ship navigators walk back into the chart house after each sight and write down the results on a work sheet. The solution I like is suitable for ship or yacht—a small vinyl board attached to a leather wrist strap. Worn on the inside of the left wrist, it can be written on while the sextant is held in the left hand. The wrist board may seem too small to write on or awkward at first, but it soon feels natural and convenient.

The pre-listing of stars to be sighted, with approximate azimuths and altitudes, including blanks for filling in *actual* altitudes and times of sights is important.

So is the order. The list should start with the star or planet that will have a usable horizon under it first.

There are a few other preparations that should be made, whether shooting the sun or stars. If using a fill-in-the-blank work sheet for reducing sights, write in the basic information before starting so it will not be necessary to stop in the midst of calculations. Neatly label all scratch paper. Make sure plotting tools are at hand, pencils sharpened, bladder emptied. Check, adjust and record sextant errors. This is the logical time to make sure mirrors and lenses are clean. Some seem afraid to touch the mirrors even to clean them and salt spray is allowed to accumulate rather than risk upsetting the delicate adjustment. This is caution to a fault. The mirrors are not that tender and any coating of film greatly reduces the quality of sights. Use disposable lens papers for this.

The single most important job: *check and record the chronometer error.*

The navigator who has done all this well in advance will have created a self-confidence building situation that minimizes doubt and improves performance, particularly as landfall approaches.

Fig. 9-4 An easily made wrist board provides a place to write sight results without putting the sextant down and keeps preplanned star location data handy. Photo by Nancy Bauer.

CHAPTER 10

SEXTANTS OF TOMORROW

So there we have the basic navigational tool kit—sextant, clock, almanac, and calculator.

Notice I include the calculator as a basic tool. The power of the modern calculator and its suitability for solving navigation problems is striking. I look on the calculator as an anti-drudgery device. Whatever brand you might choose, and there are two with optional navigational programs (Texas Instruments and Hewlett Packard) and two purely navigational calculators (Tamaya and C.Plath), you will feel liberated after having learned how to use one efficiently. I use a Hewlett Packard HP 41C with original programs developed by A.W. Fowler, a marine engineer.

These keep a running DR for an entire voyage and, in exchange for occasional insertion of data such as variation, ship's speed, air temperature, sun sights, star sights, height of eye or radio bearings will announce the ship's position at any time queried. It is a kind of continuum; corrections are applied and calculated with each new input. Such a calculator and program is the natural companion of the sextant.

What is the next step in the development of the sextant, or has it already been carried to ultimate refinement and now faces only a gentle decline into extinction as other tools relegate it to obsolescence?

I think not. An article in the June 1983 issue of *Proceedings of The Institute of Navigation*, at Washington, D.C., gives an idea of the capabilities and style of future sextants. They are

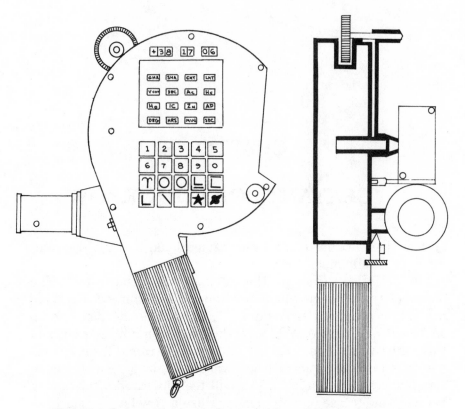

Fig. 10-1 Celnav Electronic Sextant. View right side shows keyboard for entering data identifying body being sighted. Front view shows single external mirror and index arm leading up from pivot with adjustment knob at top. Courtesy of the Institute of Navigation, Washington DC.

about as far advanced from the 1730 version of the sextant as the silicon chip is from the cat's whisker. In fact, they will use microcircuitry themselves. Called a Day/Night Sight Reduction Electronic Sextant, or Celnav, it looks something like a small movie camera with a keyboard on the right side and a scope on the left. The handle is vertical and placed under the instrument's center of gravity so that it is in balance while in a working position. This alone is a big improvement over present sextants.

The heart of the instrument is an optical drum encoder that replaces the tangent screw and limb gear teeth as well as the classic arc itself. The encoder measures the angle between the body and the horizon photoelectrically and converts the value to electrical pulses that are accepted, understood and processed by the built-in calculator. The calculator contains a quartz clock and full almanac data and when a button is pressed signaling alignment of body and horizon, calculates a LOP. After the second body has been identified and sighted, a second LOP is calculated and then a fix appears in the window expressed in terms of latitude and longitude. The fix is improved as more bodies are sighted or as more sights are taken of the same body.

Sights may be taken around the clock, because the night scope is an image intensifier, or faint light viewer, through which the navigator can see clearly by starlight. I have used one of these night viewing devices and the clarity of visibility is astonishing. When adapted to the sextant, the navigator will be able to see the horizon clearly in the dark and enjoy great flexibility in taking sights. Another important advance is achieved by the use of an averaging device so that the calculator can be fed as many as 240 altitudes and times of a single body which it then will average.

This will result in greatly improved rough weather performance.

The Celnav concept has the potential of solving all the navigator's problems, save one—overcast skies. It will be simple to get a new fix every half hour. Racing sailors could trim sails by that kind of information.

Dr. P. Kenneth Seidelman, director, *The Nautical Almanac* Office, U.S. Naval Observatory, who has contributed to the development of the design, says it is feasible to produce this instrument today, but that the price would be too high to be competitive.

Given a substantial order to spur initial production, the price would soon drop.

A pet project at C.Plath is closer to being on the counter. It is an electronic sextant that melds two proven Plath products— the Navistar Professional sextant and the Navicomp dedicated navigation calculator. I have examined the prototype instrument and found it impressive. There are two button controls for the right thumb to manipulate. One turns the computer on and off and the other directs it to accept the angle being quantified on the arc at the moment. To operate the instrument, a code to

Fig. 10-2 The pre-production C. Plath Electronic Navistar Professional Sextant. The built-in navigational computer receives time and altitude of a sight at the touch of a button on the handle. It will average several sights, a particular advantage for rough weather navigation. Price is projected at about $2,000.

specify the body being measured is keyed into the computer and the the index arm and microdrum are adjusted until the body and horizon are aligned in the optics in the usual manner. A touch of the input button causes the sight value to be entered and reduced to a LOP. Repeated sights of the same body are averaged for rough weather or poor visability conditions. Sights of several bodies will be combined for a latitude and longitude fix when the computer is keyed properly. This instrument has one advantage over the Celnav—it still can be used as a standard sextant if all the electronics should fail. It has interior batteries that are rechargeable in place.

One limitation of sextants could be overcome by applying low-level-light intensification technology. Because star sights can ordinarily be taken only during twilight, when both the horizon and star can be seen, even experienced navigators feel the pressure as the typical—and fleeting—15- to 20-minute opportunity dwindles while clouds perhaps obscure important stars. A light intensifier in place of the regular sextant scope would enable a relaxed navigator to see the horizon clearly and get fixes all night long if necessary. Fuginon Inc. sells a small, battery-powered "Starscope" that could be adapted to the sextant. It magnifies light 1,000 times, permitting a navigator to see the horizon clearly by starlight alone. Although the scope now costs about $2,300, the price might come down if sextant owners eagerly equipping their instruments generate volume sales.

These improvements may be the key to a renaissance of celestial navigation. More will come to know its charm and power. There is something perpetually awesome in the practice of consulting distant elements of the galaxy to determine location on this planet. During the course of a voyage, as the wake lengthens astern, the navigator's proficiency grows dramatically. He feels an intensifying sense of participation and incorporation in the workings of the universe. He comes to know intuitively whether a sextant sight was good or should be discarded. The triangles of his plotted fixes grow steadily smaller until they become the prized pinpoint fixes that tell the navigator his work is true. To be able to lay down these precise, single point fixes on the chart regularly day by day along the track is both the substance and reward of celestial navigation.

Appendices

Appendix A

Sextant Check Procedures

1. Set index mark 32 minutes OFF the arc.

2. Flip appropriate shades in line for both mirrors.

3. Sight on sun with sextant horizontal.

4. Touch up tangency of the two sun images with microdrum knob.

5. Record actual value of minutes OFF the arc.

6. Turn microdrum knob until the two images of the sun have exchanged places.

7. Record value of minutes ON the arc.

8. One half (1/2) difference between ON and OFF values = index error with sign direction of larger value.

9. Total of ON and OFF arc values minus index error divided by 4 = sun's semidiameter.

10. Measured semidiameter should equal the tabulated value of semidiameter found at the bottom of the sun column in *The Nautical Almanac* on the daily page for the date.

11. If the two values differ, the sextant requires adjustment. Pending adjustment, use the difference as an additional correction to Hs.

Appendix B

Sextant Manufacturers

C.Plath
Fabrik Nautischer Instrumente
Gertigstrasse 48
2000 Hamburg 60, West Germany
Phone: (040) 27191
Six basic models including three Weems & Plath

Cassens & Plath GMBH
PO Box 290 126, AM Lundeich #131
D 2850, Bremerhaven 29, Germany
Phone: (0471) 71011 TELEX: 238 609
Four basic models Cassens & Plath

B. Cooke & Son, Ltd.
Kingston Observatory
58-59 Market Place
Hull, HU1 1RH England
Phone: (0482) 223454 or 224412
The Kingston

Davis Instruments Corp.
642 143rd Ave.
San Leandro, CA 94578
Phone: (415) 483-8484
Four plastic models

EBBCO
East Berks Boat Co.
Wargrave
Berkshire, England RG 10 8JE
Two plastic models

GuangZhow Shipbuilding
Peoples Republic of China
Astra III B

Kelvin & Hughes
New Wort Rd.
Hainalt, Ilford, Essex
England IG 6 2 UR
Huson

Measure All Company
3-41 Chrome Hiroo
Shiduya-Ku
Tokyo, Japan
MAC

Observator Instrument Co.
PO Box 7155
3000 HW Rotterdam
Netherlands
MK 4

Scientific Instruments, Inc.
518 W. Cherry St.
Milwaukee, WI 53212
Phone: (414)263-1600
Mark III, Model 1

Tamaya Technics, Inc.
14-7, Ikegami 2-Chome
Ohta-Ku, Tokyo 146
Japan
Telex J-23827
Three basic models

Toizaki and Company
677-8 KA
Nagareyama - Shi Chiba-Ken
Japan
Phone: (0471) 58 4358
Junior

Appendix C

Distributors And Dealers

Mobile Instrument Repair
701 S. Conception St.
Mobile, AL 36603
(205) 438-1131

Northstar Navigation Systems
19 Caribbean Ct.
Tucson, AZ 85740

Anchorage Marine
295 Harbor Drive
Sausalito, CA 94965
(415) 332-2320

Dana Book and Navigation Co.
24402 B. Del Prado
Dana Point, CA 92629
(714) 661-3926

Davis Instruments Co.
642 143rd Ave.
San Leandro, CA 94578
(415) 483-8484

Dunne's Marine
2409 E. Harbor Blvd.
Ventura, CA 93003
(805) 644-8177

Maryland Precision Instruments
   and Optical Co.
112 E. 24th Street
Baltimore, MD 21218
(410) 467 1166

Baker Marine Instruments
2425 Shelter Island Drive
San Diego, CA 92106
(619) 222-8096

Kettenburg Marine
1880 Harbor Island Drive
San Diego, CA 92101
(619) 297-4745

Minney's Ship Chandlery
2537 West Coast Highway
Newport Beach, CA 92663
(714) 548-4192

Pacific Marine Supply
2804 Canon Street
San Diego, CA 92106
(619) 223-7194

Ship's Store
14025 Panay Way
Marina Del Rey, CA 90291
(213) 823-5574

Southwest Instrument Co.
235 W. 7th Street
San Pedro, CA 90731
(213) 519-7800

Chris Bock Instruments
13011 W. Washington Blvd.
Los Angeles, CA 90066

West Marine Products
2450 17th Avenue
Santa Cruz, CA 95062
(408) 476-1900

Great Eastern Marine
13 Hughes Circle
Ansonia, CT 06401
(203) 773-2455

Kleid Navigation, Inc.
433 Ruane St.
Fairfield, CT 06430

Mystic Seaport Museum Stores
Mystic Seaport
Mystic, CT 06337
(203) 536-9688

Bahia Mar Marine Store
801 Seabreeze Blvd.
Ft. Lauderdale, FL 33316
(305) 764-8831

Charlie's Locker
1445 SE 17th St.
Ft. Lauderdale, FL 33316
(305) 523-3350

Sailorman
350 E. State Road
Ft. Lauderdale, FL 33316
(305) 522-6718

Seabirth USA, Inc.
The Nautical Building
10 SW 6th Street
Ft. Lauderdale, FL 33318
(305) 764-8191

Navigation Equipment Co.
7750 Kenway Place E.
Boca Raton, FL 33433

Key West Marine Hardware, Inc.
818 Caroline St.
Key West, FL 33040
(305) 294-3425

Poston Marine Supply
1012 E. Cass St.
Tampa, FL 33601
(813) 229-1836

L.B. Harvey Marine
152 S.W. 8th St.
Miami, FL 33130
(305) 856-1583

Palsons
4745 NW 72nd Ave.
Miami, FL 33166
(305) 553-0450

Adler Planetarium Store
1300 S. Lakeshore Drive
Chicago, IL 60605
(312) 332-0312

Celestaire, Inc.
416 Pershing
Wichita, KS 67218
(316) 686-9785

McCurnin Nautical Charts & Co.
2318 N. Woodlawn Ave.
Metairie, LA 70001
(504) 888-4500

Cassens & Plath, U.S.
Baker Lyman & Company, Inc.
3220 S I-10 Service Road
Metairie, LA 70001
1-800-535-6956
(504) 831-3685

Robert E. White Instrument Co.
64 Commercial Wharf
Boston, MA 02110
(617) 742-3045

Hub Nautical Supply
200 High Street
Boston, MA 02110
(617) 426-9471

Better Boating Association
295 Reservoir Rd.
Needham, MA 02194
(617) 449-3314

The Hanging Locker
Mansell Road
Manset, ME 04656
(207) 244-3360

Chase Leavitt and Co.
10 Dana St.
Portland, ME 04112
(207) 772-3751

C. Plath North American Division
222 Severn Avenue
Annapolis, MD 21403
(410) 263-6700

Coast Navigation
116 Legion St.
Annapolis, MD 21401
1-800-638-0420
(410) 268-3120

Georgetown Yacht Basin
Sassafras River
Georgetown, MD 21930
(410) 648-5112

International Marine Supply
819 Pompton Ave.
Cedar Grove, NJ 07009
(201) 857-2646

Defender Industries
255 Main Street
New Rochelle, NY 10801
(914) 632-3001

Jens Jacob Nautical Service
220 Ft. Salonga Rd.
Northport, Long Island, NY 11768
(516) 757-7169

Nautical Service, Inc.
320 Rt. Salonga Rd. Rt 25A
Northport, Long Island, NY 11768

INFOCENTER Inc.
PO Box 47175
Forrestville, MD 20747
(301) 420-2468

New York Nautical Instruments
140 W. Broadway
New York, NY 10013
(212) 962-4522

Captain's Nautical Supplies
138 NW 18th Ave.
Portland, OR 97204
(503) 227-1648

Victor A. Gustin
105 S. Second St.
Philadelphia, PA 19106
(215) 922-6243

Offshore Associates, Ltd.
317 North Delaware Ave.
Philadelphia, PA 19106
(215) 592-7878

Armchair Sailor Bookstore
Lee's Wharf, Newport, RI 02840
(401) 847-4252

Coleman Supply Co.
989 Morrison Drive
Charleston, SC 29403
(803) 722-7614

Sailor's World Ship's Store
St. Thomas, U.S. Virgin Islands 08001
(809) 774-6160

Baker Lyman & Co., Inc.
602 Sawyer St.
Houston, TX 77007
(713) 864-2502

Texas Nautical Repair
2129 Westheimer St.
Houston, TX 77098
(713) 529-3551

R.H. John Chart Agency
518 23rd St.
Galveston, TX 77550
(713) 763-5742

Boat US
884 S. Pickett St.
Alexandria, VA 22304
(703) 370-4202

Backyard Boats
100 Franklin St.
Alexandria, VA 22314
(703) 548-1375

W.T. Brownley
118 W. Plume St.
Norfolk, VA 23510
(804) 622-7589

Captain's Nautical Supplies
Fisherman's Terminal
Seattle, WA 98119
(206) 283-7242

Captain's Nautical Supplies
1914 4th Ave.
Seattle, WA 98119
(206) 448-2278

Carlsen Navigation
1245 4th Ave. S., Suite E
Seattle, WA 98134
(206) 622-3433

Rex Marine Center
144 Water Street
S. Norwalk, CT 06854
(203) 866-5555

Landfall Navigation
354 W. Putnam Ave.
Greenwich, CT 06830

Bluewater B & C
1481 SE 17th Street
Ft. Lauderdale, FL 33316

Marine Navigation
615 S. La Grange Road
La Grange, IL 60525
(708) 352-0606

Fred L. Woods Nautical
76 Washington St.
Marblehead, MA 01945

Munro Sestral Ltd.
Loxford Rd.
Barking, Essex IG1 8PE
ENGLAND

Hamilton Marine
PO Box 227
Searsport, ME 04974
(207) 548-2985

Mike's Marine Supply
PO Box 191
St. Clair Shores, MI 48080
(313) 778-3200

Maryland Nautical
1143 Hull Street
Baltimore, MD 21230
(410) 752-4268

Pilothouse
1100 S. Delaware Ave.
Philadelphia, PA 19147
(215) 336-6414

Land, Sea and Sky
3110 S. Shepard Street
Houston, TX 77098
(713) 529-3551

Vega Instruments
74 Main Ave.
Bush Hill Park
Enfield, Middlesex
ENGLAND
(01) 367-0242

Appendix D

The Navigator's Basic Tool Kit

NOTEBOOK—It is important to make navigation into a routine as completely as possible. Nothing helps more than a record of what has been done before, day by day.

The easiest way to do this is with a notebook with a space for each day's calculations. Arrangement should be such that is is easy to compare today's figures with those of the previous day.

A standard format is important.

WORK SHEETS—These are effective aids for organizing routine. Davis Instruments prints sheets for each method. *Bowditch* gives models that can be copied and used as a guide while working in the notebook. Dunlap's *Celestial Navigation with HO 249* also includes some fine forms for use with that publication.

PLOTTING SHEETS—For laying out the results of sights at a larger scale and without having to use and abuse the chart itself. The kind cailed Universal Plotting Sheets are cheap and handy. They cost about $3.50 for 50 from the Defense Mapping Agency (see Appendix G).

PLOTTING INSTRUMENTS—Dividers for measuring distances and transfering them from place to place are a necessity. Cheap dividers that slip are a curse. The British one-handed dividers with the flat loop at the top and blunt points are incompatible with accurate measurment.

Ordinary dividers can be handled with one hand quite well.

PARALLEL RULES—The simplest means of transfering a line parallel to another is done with the parallel rule. Of the several versions I have seen and used, the very best seems to be the type that is double jointed with an extra bar in the center to ensure the two edges maintain their parallelism while moving straight apart. Ordinary parallel rules open out on a diagonal and must be walked along in an awkward fashion.

TRIANGLE—Many navigators use large triangles for transfering parallel lines. If you don't use triangles for parallel lines, you still need at least one for plotting the line of position perpendicular to the azimuth of the body sighted. The smaller the handier. It also makes a good edge for holding under a line of figures in *The Nautical Almanac*.

COMPASS—A good long-legged draftsman's compass with a properly sharpened point is more useful than dividers for much chart work. When laying off a course or an intercept, it is handy to be able to make a pencil mark quickly at the point measured. You can punch a hole in the chart with the dividers, but you still must pick up a pencil to mark the spot.

MAGIC TAPE—Softer and thinner than the old sticky Scotch Tape, Magic Tape is wonderfully transparent when stuck to a chart. Importantly, it can be written on with a pen or pencil and, in the latter case, erased clean again and again.

CHARTS—The importance of having a good chart at the right time cannot be overstated. It is particularly difficult to make sense of radio direction finder data without a regular chart of the area that extends well offshore. It stocking charts for a voyage, consideration should be given to the possibility that one may have to put into ports along the way for emergency reasons. Merchant marine ships are a good source of used charts. They are constantly replacing charts with the newest issues and discarding slightly outdated ones.

Appendix E

Making And Using
An Artificial Horizon

An artificial horizon is effective on land and easy to make and use. It is possible to get sights that are more accurate than sights at sea, though they are not quite good enough for surveying purposes.

To make one, place a pie tin or other shallow flat container level on the ground and pour enough oil in the bottom to cover the surface completely. New or used engine oil will do as will cooking oil. If the oil is too clear, add dark food coloring to make a good reflecting surface. Try to do this in a windless spot as ripples will cause the exercise to be worthless.

Once the oil is in place on the ground between you and the sun, back off until the sun's reflection can be seen in the oil. Then, with the sextant at zero, and with the appropriate shades in place, sight on the reflection and move the index arm out slowly until another sun appears in view (this is one time when it is easy to bring down a body without losing it). Keeping the oil reflection in view keeps the instrument aligned on the correct azimuth during the process. The image of the sun brought down is double reflected, therefore erect, which is to say that what looks like the lower limb is the lower limb. Match the lower limb to the top of the oil-reflected image until the two are just touching. Divide the angle by two to get sextant altitude (Hs).

The sight must be adjusted with all the usual corrections except for height of eye.

If the sun's altitude is above 30 degrees (where refraction is minimal), index error can be accurately determined by taking readings of both upper and lower limbs of the sun and subtracting the lessor figure to get the diameter of the sun. Half of that result compared with *The Nautical Almanac* semidiameter value at the bottom of the daily page will yield the index error. The index correction must be applied to the angle measured with the artificial horizon before it is divided by two; all other corrections are made later.

Appendix F

Table of Interstellar Angles
For Practice Sighting and Sextant Testing

EXPLANATION:

With the sextant set at 0 (zero), sight on the lower or left-most star of the pair.

Slide index arm out slowly until the second star comes into view. Using the microdrum adjustment, *precisely match the two stars*. Smaller angles are easier to measure as are stars in a vertical plane. Star pairs in the horizontal plane give more accurate results because when altitudes are about the same, refraction can be ignored.

When measuring the two stars in a vertical plane, use corrections taken from *The Nautical Almanac,* Table A. Since refraction causes stars to appear higher than they are, the difference between the refraction corrections for the upper and lower stars must be added to the sextant reading to get the correct interstellar angle, e.g., upper star at an altitude of about 60 degrees calls for a correction of -0.6; the lower star at about 15 degrees is corrected by -3.6. The combined correction is 3 minutes, which is to be added to the sextant reading. Index error must be applied in the regular fashion, whether the star is vertical or horizontal.

Interstellar distances do fluctuate slightly over time. For example, the distance between *Aldebaran* and *Alnilam* computed in 1895 by Capt. Squire Lecky, for a table in his *Wrinkles in Practical Navigation*, was 23 degrees –08'. Calculated from 1985 *Nautical Almanac* data the value was 23 degrees –7.9', and by 1991 data, it is back to 23 degrees –08'. Distances between other stars may vary more over shorter cycles.

Values shown in this table can be calculated afresh using the formula for Great Circle distance found in *Bowditch*. SHA and dec. for each star are substituted for the latitudes and longitudes of the points of departure and destination. The resultant distance in miles is divided by 60 to produce degrees and minutes.

Table of Interstellar Angles
For Practice Sighting and Sextant Testing

	Procyon	Betelgeuse	Bellatrix	Aldebaran	Pollux	Capella	Rigel
Sirius	25° –42.1'	27° –06.2'	30° –23.2'	46° –01.4'	47° –03.1'	65° –49.6'	23° –40.4'
Rigel	38° –30.8'	18° –36.4'	14° –47.3'	26° –29.7'	51° –22.2'	54° –12.2'	—
Betelgeuse	25° –57.8'	—	7° –31.9'	21° –23.4'	33° –12.0'	39° –28.4'	18° –36.4'
Bellatrix	33° –23.3'	7° –31.9'	—	15° –45.3'	39° –37.7'	39° –41.5'	14° –47.3'

	Arcturus	Vega	Deneb	Alkaid	Kochab	Schedar	Regulus
Spica	32° –47.7'	87° –46.9'	—	60° –40.2'	86° –22.2'	—	54° –03.4'
Altair	81° –15.9'	34° –11.8'	38° –01.1'	83° –48.7'	77° –24.5'	72° –57.8'	—
Dubhe	53° –55.9'	66° –03.3'	69° –12.1'	25° –42.4'	23° –19.2'	60° –12.5'	50° –46.9'
Vega	59° –06.8'	—	23° –50.9'	51° –01.1'	43° –57.4'	58° –56.9'	—

Data (SHA & dec.) from 1991 *Nautical Almanac*.

Appendix G

Useful Addresses

Defense Mapping Agency
Office of Distribution Services
Attention: DDCP
Washington, D.C. 20315
Customer Service Telephone: (202) 227-2816
Source of government charts and publications. Send for "Public Sale Catalog."

Navigator Publishing Corp.
18 Danforth Street
Portland, ME 04101
(207) 772-2466
Publishers of "Ocean Navigator," a bimonthly magazine devoted to nagivation.

Foundation for The Promotion of The Art of Navigation
P.O. Box 1126
Rockville, Maryland 20850
(301) 622-6448
Publishers of "Navigator's Newsletter."

The Institute of Navigation
815 15th Street, Suite 832
Washington, D.C. 20005
Phone: (202) 783-4121
An organization that promotes the science of navigation.

William F. Nye, Inc.
P.O. Box G-927
New Bedford, MA 02742
Phone: (617) 996-6721
Telex: 940 807
Source of 140B Clock Oil described in Chapter 4.

Jerry Co.
American Science Center
601 Linden Place
Evanston, Illinois 60202
Phone: (312) 475-8440
Source of front-surface, high quality mirror stock suitable for sextant use.
Catalog on request.

Maryland Precision Instrument and Optical Co.
Frank J. Janicek, Pres.
112 E. 24th Street
Baltimore, Maryland 21218
Phone: (410) 467-1166
Sextant repair and adjustment, mirror resilvering

Captain's Nautical Supplies
Bill Cook, Manager,
   Precision Instruments and Optics Div.
Fishermen's Terminal
Seattle, WA 98119
(206) 283-7242 or 1-800-448-2278
Chromed front-surface mirrors, repairs, collimation,
reconditioning, and reconditioned sextant sales.

Bibliography

Anon., *Publication 117, Radio Navigational Aids.* 2 vols. Washington, DC: Defense Mapping Agency Hydrographic Center, 1984.

Birney, Arthur A., *Sun Sight Navigation: Celestial for Sailors.* Centerville, Maryland: Cornell Maritime Press, 1984.

Birney, Arthur A., *Noon Sight Navigation.* Centerville, Maryland: Cornell Maritime Press, 1979.

Blewitt, Mary, *Celestial Navigation for Yachtsmen.* New York: John de Graff, Inc., 1981.

Boorstin, Daniel J., *The Discoverers.* New York: Random House, Inc., 1983.

Bowditch, Nathaniel, *American Practical Navigator: An Epitome of Navigation.*
    71st ed., Vol. 1. Washington, DC: Defense Mapping Agency Hydrographic Center, 1984.

Brown, Otis S., *One Day Celestial Navigation for Offshore Sailing.* Greenbelt, Maryland: C&O Research Co., 1984.

Bowen, Catherine Drinker, *The Most Dangerous Man in America.* New York: Little, Brown & Co., 1974.

Budlong, J.P., *Sky And Sextant: Practical Celestial Navigation.* New York: Van Nostrand Renhold Co., 1975.

Cotter, Charles H., *A History of the Navigator's Sextant.* Glasgow: Brown, Son & Ferguson, 1983.

Crawford, William P., *Mariners Celestial Navigation.*
    New York: W W Norton and Co., Inc., 1979.

Dahl, Norman, *The Yacht Navigator's Handbook*. New York: Hearst Marine Books, 1983.

Dunlap, G.D., *Successful Celestial Navigation with H.O. 229*. Camden, Maine: International Marine Publishing Co., 1977.

Dunlap, G.D., and H.H. Shufeldt, *The Book of the Sextant*. 4th ed. Annapolis, Maryland: Weems and Plath, 1980.

Evans, John, *Once-A-Day Navigation—Three Easy Methods to Find Latitude And Longitude*. San Diego: Polaris Marine Services, 1984.

Fraser, Bruce, *Weekend Navigator*. New York: John De Graff, Inc., 1981.

Hart, M.R., *How to Navigate Today: A Straightforward Guide to Practical Navigation*. Cambridge, Maryland: Cornell Maritime Press, 1970.

Kleid, Robert E., *Choosing A Marine Sextant*. 4th ed. Fairfield, Connecticut: R.E. Kleid, 1975.

Lecky, Squire Thornton Stratford, *Wrinkles in Practical Navigation*. 22nd ed. Princeton: D. Van Nostrand, 1937.

Letcher, John S. Jr., *Self Contained Celestial Navigation with H.O. 208*. Camden, Maine: International Marine Publishing Co., 1977.

Maloney, Elbert, *All About Sextants*. Annapolis, Maryland: Coast Navigation, 1978.

Mixter, George W., *Primer of Navigation*. 5th Edition. Princeton: D. Van Nostrand, 1967.

Moody, Alton B., *Navigation Afloat—A Manual for Seamen*. New York: Van Nostrand Reinhold Co., 1980.

Norville, Warren, *Celestial Navigation Step by Step*. Camden, Maine: International Marine Publishing Co., 1981.

Rantzen, M.J., *Little Ship Astro-Navigation*.
   4th ed. London: Barrie and Jenkins, 1977.

Schelereth, Hewitt, *Celestial Navigation by Star Sights*. Newport,
Rhode Island: Seven Seas Press, 1983.

Seaton, S.L., *Sun Sight Sailing: Foolproof Step by Step Method of Finding Your Position at Sea*. New York: David McKay Co., 1980.

Shufeldt, Henry H., and Newcomer, Kenneth E., *The Calculator Afloat*.
   Annapolis, Maryland: U.S. Naval Institute Press, 1980.

Thompson, Joseph E., *Celestial Navigation: Capt. Joe Thompson's
Cookbook Method*. New York: David McKay Co., 1981.

Van Doren, Carl, *Benjamin Franklin*. New York: Viking Press, 1938.

Index

Adjusting (Adjustment) Screw, 42, 52, 55, 71, 78, 85
Altair, 151
Altitude, 17, 18, 20, 22, 23, 62, 103, 104, 105-110, 113, 115, 116, 117, 126, 131, 132, 134, 139, 150, 151, 152, 154, 159
Anderton, Paul, 47, 80
Apparent Altitude (Ha), 123, 130, 132, 134
Arc, 18, 20, 22, 23, 32, 33, 40, 41, 42, 43, 46, 47, 48, 50, 52, 60, 62, 64, 69, 70, 71, 87, 93, 105-110, 123, 126, 127, 134, 142, 158, 159
Aries, 148, 149, 152
Artificial Horizon, 88
Astrolabe, 18, 22
Astigmatizer, 57, 58, 59
Azimuth, 103, 104, 105, 139, 151, 152, 154
Backstaff, 21, 23, 25, 31
Baker, Lyman & Co., 93
Bauer, Cmdr. Bruce 33
Big Dipper, 58
Black Mirror Correction (BMC), 82
Boorstin, Daniel, 28
Bowditch, Nathaniel, 20, 93, 109
Brandies and Sons, 45
Bubble, 63
Bubble Sextant, 45, 62
Budlong, J.P., 78
Cassens & Plath, 34, 57, 58, 60, 62, 78, 84, 92, 93, 98, 125, 145
Carver, 93
Celnav, 158, 159, 160
Circular Collar, 43
Clamp Release, 41
Clock Oil 140-B, 70
Collimation Error, 52, 54
Columbus, 20
Computed Altitude (Hc), 123, 152

C. Plath, 42, 60, 62, 84, 90, 91, 92, 93, 157, 159
Cranmer, J.A., 89
Cross-Staff, 20, 23, 24
Davis, Col. Warren P., 18, 21
Davis Instruments, 58, 60, 61, 62, 84, 106
Davis, John, 21, 22, 23
Davis Prism, 60, 61, 62, 106
Dip (D), 127, 128, 132, 134, 135
Double Reflection (Reflecting), 25, 26, 28, 30, 31, 32, 35
Dunlap, G.D., 45, 46, 66
Einstein, 26
Eyepiece, 54, 119, 121
Fornili, Dr. Robert, 66
Foulkes, Thomas, 88, 89, 98
Fowler, Alex, 142, 157
Frame, 32, 37, 42, 43, 45, 47, 48, 52, 54, 62, 64, 71, 72, 73, 78,
    91, 93, 101, 106, 144, 145
Franklin, Benjamin, 30
Freiberger, 60, 143
Fulvew Horizon Glass, 44
Galileo, 110
Gear Rack, 40
Gear Teeth, 37, 69, 158
Godfrey, Thomas, 29, 30, 31, 34
Greenwich Hour Angle (GHA), 134, 148, 149
Greenwich Mean Time (GMT), 140, 142, 149
Gyro Compass, 103, 104
Hack Watch, 137, 140
Hadley, John, 26, 28, 30, 31, 32, 34, 62
Halley, Edmund, 26, 28, 30
Handle, 43, 45, 56, 64, 71, 72, 85, 101, 105, 142, 143, 145
Height Of Eye, 117, 127, 128, 135, 157
Heuer Microsplit 1000, 145
Hewlett Packard, 157
Hiscock, Eric, 57, 116
HO 229, 66, 117
HO 249, 103, 105, 148, 152, 153
Hooke, Dr. Robert, 26
Horizon Glass, 31, 34, 42, 43, 45, 48, 50, 52, 55, 58, 60, 62, 78,
    80, 81, 85, 89, 91, 107, 110, 111, 112, 113, 127, 144
Horizon, Artificial, 88
Horizon Mirror, 52, 54, 60, 110

Index Arm, 33, 40, 41, 42, 46, 47, 48, 54, 69, 71, 93, 101, 103, 142, 160
Index Correction (IC), 82, 113, 127, 134
Index Error, 45, 50, 52, 78, 85, 87, 113, 127
Index Mark, 40, 50
Index Mirror, 33, 42, 45-48, 52, 55, 57, 58, 60, 75, 78, 85, 91, 106, 110-114, 127, 144
Instrument Error (I), 123, 134
Irradiation, 108, 109, 113
Jojoba Oil, 70
Kamal, 17, 19, 20
Kliemann, Dr. N, 91
Kochab, 23
Latitude, 17, 18, 21, 23, 86, 110, 150, 152, 159, 160
Latitude Hook, 17, 19
Lecky, Squire, 69
Leg, 45, 64, 72, 73, 75, 76, 102
Letcher, John S., 83
Limb, 37, 40, 69, 70, 71, 72, 73, 75, 76, 123
Litton, 90, 91
Local Apparent Noon (LAN) 64, 86, 108, 147
Local Hour Angle (LHA) 148, 149, 150, 152, 153
Logan, John, 30
Longitude, 26, 29, 30, 149, 159, 160
Loran, 92
Lower Limb, 96, 109, 110, 132
Mars, 134
Maryland Precision Instruments, 79
Maskelyne, Rev. Dr. Nevil, 79
Micrometer Drum (Microdrum), 42, 43, 48, 84, 87, 89, 93, 103, 107, 113, 127, 129, 160
Micrometer Scale, 41
Monocle, 119, 120, 121
Nautical Almanac, The, 23, 42, 79, 82, 86, 87, 103, 109, 128, 130, 132, 134, 137, 147, 148, 149, 159, 161
Navicomp, 159
Navistar Professional, 159, 160
Navy Mark III, 54, 60, 78
Neck Strap, 64, 65
Newton, Sir Isaac, 25, 26, 28, 29, 31
Nocturnal, 23, 25, 27
North Star, 17, 18
Nye, William F., 70

Observed Altitude (Ho), 123, 134
Octant 32
Oldenburg, Henry, 28
Omega, 92
Pactra, 80
Paint, 79, 80, 88
Parallelism, 45, 50, 52
Parallel Mirrors, 33
Peary, Robert Edwin, 101
Pelorus, 104
Perpendicularity, 45, 47, 48, 49, 50
Pivot, 40
Pivot Point, 18
Plotting Sheet, 104
Polaris, 17, 23, 25,
Prism Level, 60, 61, 106
Quadrant, 18, 20, 23, 30
Rantzen, M.J., 117
Refraction, 82, 87, 108, 132, 134
Resilvering, 79, 85, 87
Rising Piece, 43
Royal Society, The, 25, 26, 28, 30, 31
Rude, Capt. Gilbert T., 148
Rude Star Finder, 148, 150, 151, 152
SatNav, 92
Scope, 43, 48, 54, 55, 58, 60, 62, 64, 71, 75, 76, 79, 95, 96, 101, 102, 103, 109, 118, 119, 159
Semi-Diameter (SD) 82, 87
Seidelman, Dr. P. Kenneth, 159
Sextant Altitude (Hs), 123, 130, 134, 139, 142
Sextant Hook, 73, 74, 102
Shade Glasses (Shades), 42, 71, 75, 110-114, 122
Side Error, 48, 50, 51, 52, 53, 78
Silvering, 42, 79, 80, 85
Simex, 60
Simmondsia Chinensia, 70
Sirius, 96, 113
Specula, 79
Split-Hand Stopwatch, 140, 141
Star Scope, 43, 95
Stellar Lenticular, 57
Sun Shade, 57

Tamaya, 60, 62, 84, 90, 93, 157
Tangent Worm Screw (Tangent Screw), 34, 40, 42, 50, 52, 69, 71, 84, 89, 103, 158
Telescope, 42, 52, 54, 71, 101, 102, 104, 110, 111, 119, 121
Texas Instruments, 157
Time Tick, 139, 140, 142
Timex, 143, 145
Timing Watch, 43
Universal Time (UT), 149
Upper Limb, 108, 109, 110, 132, 134
Vega, 151
Vega Instruments, 143
Venus, 64, 95, 103, 113, 114, 134
Vernier (Scale), 41, 42, 69, 84
Watts, Capt. Oswald M., 85
WD-40, 71, 72
Weems And Plath, 60
Whiteside, Dr. D.T., 31
Wide View Horizon Glass, 51, 58, 60
Wollaston Double Star Prism, 58
Wrist Board, 154, 155
Zenith (Zn), 18, 62, 105, 152

Notes

Notes

10/94 5
4/60 18
8/05 21